T0331016

Multiple Factor Analysis by Example Using R

Chapman & Hall/CRC
The R Series

Series Editors

Aims and Scope

This book series reflects the recent rapid growth in the development and application of R, the programming language and software environment for statistical computing and graphics. R is now widely used in academic research, education, and industry. It is constantly growing, with new versions of the core software released regularly and more than 5,000 packages available. It is difficult for the documentation to keep pace with the expansion of the software, and this vital book series provides a forum for the publication of books covering many aspects of the development and application of R.

The scope of the series is wide, covering three main threads:

- Applications of R to specific disciplines such as biology, epidemiology, genetics, engineering, finance, and the social sciences.
- Using R for the study of topics of statistical methodology, such as linear and mixed modeling, time series, Bayesian methods, and missing data.
- The development of R, including programming, building packages, and graphics.

The books will appeal to programmers and developers of R software, as well as applied statisticians and data analysts in many fields. The books will feature detailed worked examples and R code fully integrated into the text, ensuring their usefulness to researchers, practitioners and students.

Published Titles

Using R for Numerical Analysis in Science and Engineering , *Victor A. Bloomfield*

Event History Analysis with R, *Göran Broström*

Computational Actuarial Science with R, *Arthur Charpentier*

Statistical Computing in C++ and R, *Randall L. Eubank and Ana Kupresanin*

Reproducible Research with R and RStudio, *Christopher Gandrud*

Introduction to Scientific Programming and Simulation Using R, Second Edition, *Owen Jones, Robert Maillardet, and Andrew Robinson*

Displaying Time Series, Spatial, and Space-Time Data with R, *Oscar Perpiñán Lamigueiro*

Programming Graphical User Interfaces with R, *Michael F. Lawrence and John Verzani*

Analyzing Baseball Data with R, *Max Marchi and Jim Albert*

Growth Curve Analysis and Visualization Using R, *Daniel Mirman*

R Graphics, Second Edition, *Paul Murrell*

Multiple Factor Analysis by Example Using R, *Jérôme Pagès*

Customer and Business Analytics: Applied Data Mining for Business Decision Making Using R, *Daniel S. Putler and Robert E. Krider*

Implementing Reproducible Research, *Victoria Stodden, Friedrich Leisch, and Roger D. Peng*

Dynamic Documents with R and knitr, *Yihui Xie*

Multiple Factor Analysis by Example Using R

Jérôme Pagès
Agrocampus-Ouest
Rennes, France

CRC Press is an imprint of the
Taylor & Francis Group, an **informa** business
A CHAPMAN & HALL BOOK

CRC Press
Taylor & Francis Group
6000 Broken Sound Parkway NW, Suite 300
Boca Raton, FL 33487-2742

ISBN-13: 978-1-4822-0547-3 (hbk)

Visit the Taylor & Francis Web site at
http://www.taylorandfrancis.com

and the CRC Press Web site at
http://www.crcpress.com

Contents

Preface

Due to the extensive field of application, multiple factor analysis (MFA) is widely used today. This book is the first comprehensive account of the method in English: it brings together the theoretical and methodological aspects with application examples and details of how to implement them using an R package (**FactoMineR**).

In the same way as for principal component analysis (PCA) or multiple correspondence analysis (MCA), MFA is applied to tables in which a set of individuals (one individual = one row) is described by a set of variables (one variable = one column). The particularity of MFA lies in the fact that within the active variables, it can account for a group structure defined by the user. Such data tables are called *individuals × variables organised into groups*.

This data format is widely used, firstly because it corresponds to the user's approach when designing data collection. For example, someone designing an opinion poll organises the questionnaire into themes, each of which is then developed using several questions (the questions are the variables and the themes the groups of variables). This structure must of course be present when analysing the results of the poll. Secondly, it is widespread because users often want to bring together data collected for the same statistical individuals but in different contexts (for example, geographical or temporal). Thus, in the food industry, for a given set of products, we often have sensory profiles from tastings conducted in different countries. These sets of data need to be analysed simultaneously while at the same time preserving their individuality both during the statistical analysis and the interpretation phase.

Experience in working with many diverse users has shown that multiple tables are in fact the standard data format used today. Along with this complex structure (into groups of variables), the nature of the data is also complex as the variables can be quantitative or qualitative. It is therefore necessary for users to have access to a methodology for analysing individuals × variables tables in which the variables are structured into quantitative, qualitative or mixed groups. This is precisely the field of application of MFA.

MFA is the result of joint research by Brigitte Escofier and Jérôme Pagès in the early 1980s. This method is now well established if we consider the wide range of software available. To name but a few software packages including an MFA procedure: **FactoMiner** (R Package) **ade4** (R Package), SPAD, Uniwin (Statgraphics) and XLStat.

Having achieved widespread availability of the method, and with the data format justifying its implementation, one question remains: what exactly does 'account for a group structure of the variables in an overall analysis' mean? In other words, why do we not simply conduct, for example, a principal component analysis and take the group structure of the variables into account solely during interpretation? We might summarise by saying that this book answers this question first and foremost.

The first two chapters look back at basic factorial analysis methods for individuals × variables tables, PCA and MCA.

Chapter 3 presents Factor Analysis of Mixed Data (FAMD), a little-known method for simultaneously analysing quantitative and qualitative variables without group distinction. FAMD contains the technical elements required for taking into account both types of variables within one single analysis.

The following chapters, numbered 4 to 9, describe multiple factor analysis. The first four look in turn at the key points of MFA in the context of quantitative variables. In addition, one chapter is given over to qualitative and mixed data. Finally, one chapter compares MFA and Procrustes analysis.

Chapter 10 presents a natural extension of MFA: hierarchical MFA (HMFA). In this method, the variables are not structured by a simple partition, but by a hierarchy. A typical example of these data is surveys, for which the questionnaire is organised into themes and subthemes.

The final chapter presents several elements of matrix calculation and metric spaces used in the book in the form of two technical appendices.

***

To conclude this work, it gives me great pleasure to thank Sophie Puyo and Magalie Houée-Bigot, the statistical engineers who were responsible for the majority of the layout for this book. I also thank Eric Matzner-Løber, editor for the French version of this book for his contributions. Thanks also go to Rob Calver at Chapman & Hall for his friendly assistance. A special mention goes to Rebecca Clayton for her invaluable help in translating from the French to the English version. I would also finally like to thank my wife Annie who brightens my life and therefore, indirectly, this book.

***

The data and the R scripts used in this book are available on the website of the applied mathematics department at Agrocampus Ouest.

Chapters 3, 8 and 9 are adapted from works first published in the *Revue de Statistique Appliquée* (*Journal of Applied Statistics*, which ceased publication in 2006). This is an excellent occasion to thank Pierre Cazes, director of the journal, first for his enthusiastic reception of the work and then for his continued encouragement concerning the adaptation of the book.

Thanks also go to Richard Delécolle and his talent for calligraphy.

1

Principal Component Analysis

Principal component analysis (PCA) is the most widely used factorial method. It is applied to tables in which a set of (statistical) individuals is described by a set of quantitative variables. In this chapter, we present a detailed description of this method, both in theory and in practice. This is the perfect opportunity to introduce a number of concepts used to analyse multiple tables, but also apply to simple ones. This enables the reader to see the specificities of multiple factor analysis (MFA) better.

Vocabulary: Factor Analysis or Factorial Analysis?

Both families of methods are very similar, which explains the confusion between the two names. Roughly, we can say that *factor analysis* is based on a model whereas *factorial analysis* is purely exploratory. Typically, *principal component analysis* is a factorial analysis. The same is true of correspondence analysis and all the methods presented in this book. In this case, why do we say *multiple factor analysis*? It is simply a mistranslation of a method originally introduced in French under the name *Analyse factorielle Multiple*. Now, *multiple factor analysis* is widespread and, in our opinion, changing its name to *multiple factorial analysis* would lead to yet more confusion.

1.1 Data, Notations

We start by studying a table with the following characteristics:

- Each row represents a statistical individual; we denote I the number of individuals. I also designates the set of individuals. Using the same letter to designate both the set and its cardinal is not confusing as the meaning is always clear from the context.
- Each column represents a quantitative variable; K represents the number of variables (as well as the set of variables).
- At the intersection of row i and column k, is x_{ik}, the (numerical) value of individual i for variable k.

Let us add two classical notations:

$\bar{x}_k$: Mean of variable k; this is not used much as the variables are generally centred, but it can sometimes be useful for the centring to appear explicitly;

s_k: The standard deviation of variable k.

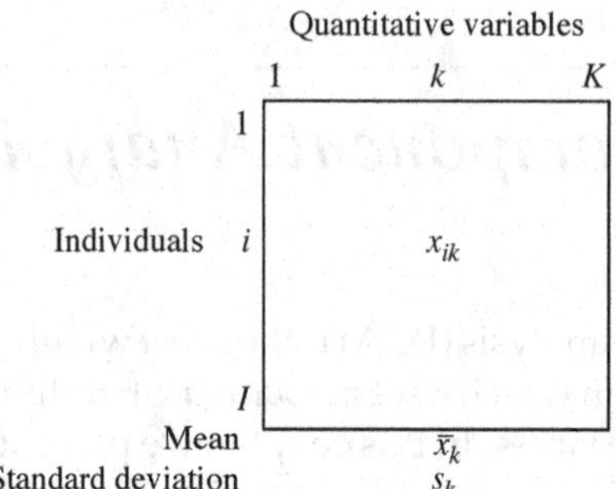

FIGURE 1.1
Data structure and notations.

These notations are brought together in Figure 1.1

An endless variety of data can be analysed using PCA. In the following paragraphs, we use an example which is both rich and easy to understand. We have the final Baccalaureate grades for 909 high school students specialising in science ($I = 909$), for 5 given subjects ($K = 5$): mathematics, physics, natural sciences, history–geography and philosophy.

1.2 Why Analyse a Table with PCA?

Let us go back to the previous example. Generally, once the means have been examined, the aim of studying such a table statistically is to investigate the students' 'within-subject' and 'between-subjects' diversity. This diversity must first be examined by subject using indicators (mainly standard deviations) and graphs (mainly boxplots and histograms).

The choice of PCA is mainly motivated by two objectives:

1. We consider the students not in terms of one specific grade or another, but in terms of overall grades, which we refer to as their 'school profile.' We then study the diversity of these profiles (overall, rather than grade by grade). In PCA, this profile diversity is studied by highlighting their principal dimensions of variability. Therefore, in the example, we can expect the principal dimension of variability to oppose good students (those who have good grades in all subjects) with bad students (those who have bad grades in all subjects).
2. We are interested in the relationships between variables. In PCA, we only examine linear relationships; the intensity of this type of relationship between two variables is measured by the correlation coefficient, as usual. In addition, these relationships are studied using synthetic variables (known as principal components). These are

linear combinations of initial variables which are as closely related as possible (defined later) to these initial variables. Ideally, each synthetic variable is closely correlated to one group of variables alone, and uncorrelated with the others, thus identifying groups of variables (correlated within-group and uncorrelated between-groups).

We show that these synthetic variables coincide (defined later) with the dimensions of variability as seen above. This therefore shows that the two objectives are closely linked, or are even two aspects of the same question. This can be illustrated in the context of the example: saying that the principal dimension of variability opposes good and bad students (studying the individuals via their school profiles) is equivalent to saying that all of the variables (that is, the grades) are positively correlated two by two (studying relationships between variables).

This idea may seem obvious: the rows and the columns of a table are two aspects of the same reality (that is to say the table itself). This is the origin of the term *duality* (the dual nature) often used to designate this relationship between these two aspects, on the one hand, and between the coordinates of rows and columns given by PCA. This relationship is no less fundamental: it helps us to understand better what it is we're looking for; it also illustrates the suitability of the PCA for a very general issue, such as analysing a table. In addition, it must be noted that we also find this duality (of issues and results) in all factorial analyses, in particular those studied in this book, PCA, multiple correspondence analysis (MCA), factorial analysis of mixed data (FAMD), MFA and hierarchical multiple factor analysis (HMFA).

1.3 Clouds of Individuals and Variables

Cloud of Individuals N_I

We associate individual i with its profile $\{x_{ik}; k = 1, K\}$. Point M_i corresponds to this profile in $\mathbb{R}^K$, in which each dimension represents a variable (see Figure 1.2). $\mathbb{R}^K$ is said to be the *individuals' space*. Set I of points i makes up a cloud denoted N_I. In addition, each individual is attributed the weight p_i so $\sum_i p_i = 1$ (generally $p_i = 1/I$).

The centre of gravity for cloud N_I (denoted G_I, also known as the *mean point*), has the coordinates $\{\bar{x}_k; k = 1, K\}$. When the variables are centred, as is always the case in PCA, the origin of the axes in $\mathbb{R}^K$ is positioned in G_I (additional information on centring data is given in Section 1.4).

In cloud N_I, the squared distance between two individuals, i and l, is expressed:

$$d^2(i,l) = \sum_k (x_{ik} - x_{lk})^2.$$

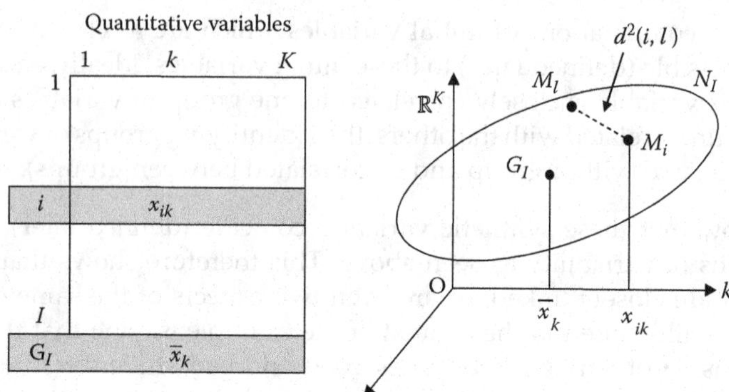

FIGURE 1.2
The cloud of individuals.

This quantity measures the discrepancy between the profiles of individuals i and l. Studying the individuals' variability means studying these distances, the whole set of which determines the shape of cloud N_I. This variability can also be dealt with using the distances between each point M_i and the mean point G_I. Therefore, for individual i:

$$d^2(i, G_I) = \sum_k (x_{ik} - \bar{x}_k)^2.$$

This distance measures the 'peculiarity' of the individual i (how it differs from others). The set of these individual peculiarities makes up the overall variability of the data. To measure this overall variability, the squares of the distances to the mean point are aggregated in order to obtain the total inertia of N_I (with respect to G_I). Thus

$$\text{Total inertia of } N_I/G_I = \sum_i p_i d^2(i, G_I) = \sum_k \sum_i p_i (x_{ik} - \bar{x}_k)^2 = \sum_k \text{Var}[k].$$

This total inertia is equal to the sum of the K variances, denoted Var[k]. Thereby, when the variables are reduced, it is equal to the number of variables. This clearly shows, in the centred-reduced case but also in general, that, in PCA, it is not the total inertia itself which is interesting but rather the way in which it is distributed. We also observe this property in MCA and MFA. We obtain the same total inertia in aggregating the squares of the between-individuals distances, a perspective used at the beginning of this section. The variance of variable k, based on the deviations between individuals, is thus expressed:

$$\text{Var}[k] = \frac{1}{2} \sum_i \sum_l p_i p_l (x_{ik} - x_{lk})^2.$$

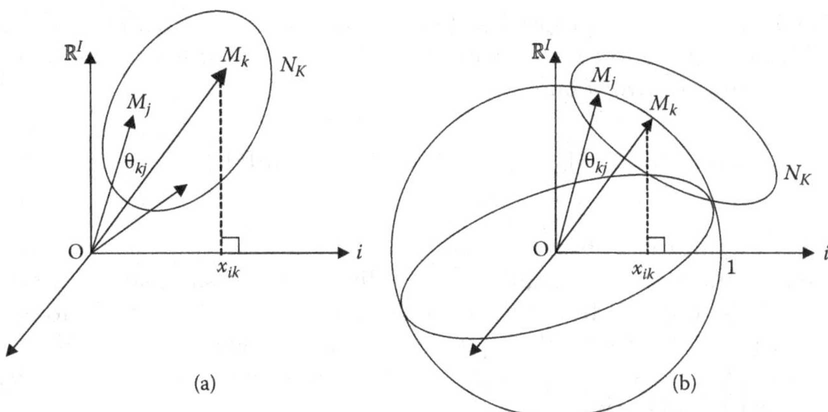

FIGURE 1.3
The cloud of variables. A: centred data; B: centred and reduced data. θ_{kj} is the angle formed by the two vectors representing variables k and j ($\overrightarrow{OM_k}$ and $\overrightarrow{OM_j}$).

By combining the two previous equations, we obtain:

$$\text{Total inertia of } N_I/G_I = \frac{1}{2}\sum_{i,l} p_i p_l \sum_k (x_{ik} - x_{lk})^2 = \frac{1}{2}\sum_{i,l} p_i p_l d^2(i,l).$$

This shows that the inertia of N_I represents the individuals' variability both from the point of view of their deviation from the centre of gravity and from the point of view of the between-individuals distances.

Cloud of Variables N_K

To variable k, we attribute its values for all of the individuals studied $\{x_{ik}; i = 1, I\}$. This set corresponds to point M_k (and to vector v_k) in space $\mathbb{R}^I$, in which each dimension corresponds to an individual. $\mathbb{R}^I$ is called the *variables' space* or, more generally, the *space of functions on I* (a function on I attributes a numerical value to each individual i). The set of points M_k constitutes the cloud of variables denoted N_K (see Figure 1.3).

When the variables are centred, as is always the case in PCA, this space has two remarkable properties.

1. The cosine of the angle θ_{kj} formed by the two variables k and j is equal to their correlation coefficient. This geometrical interpretation of the correlation coefficient justifies the use of this space to study the relationships between variables. It also explains that we represent variable k by the vector linking the origin to point M_k.
2. The distance between M_k and O is equal to the variance of variable k. Interpreting a variance as a squared length is extremely valuable in statistics. Note: a centred-reduced variable has a length of 1; cloud N_K is then situated on a hypersphere (with a radius of 1).

To obtain these two properties, it is necessary, when calculating a distance in $\mathbb{R}^I$, to attribute to each dimension i the weight p_i of the corresponding individual. Thus, we obtain:

$$d^2(O, M_k) = \sum_i p_i\,(x_{ik} - \bar{x}_k)^2 = \mathrm{Var}[k].$$

This way of calculating the distance (also known as the *metric)* in $\mathbb{R}^I$ is therefore linked to the weights of the individuals. We say that the weights on the individuals induce a metric in the space of functions on I. To get a feel for this, we consider the equivalence between two rigorously identical individuals of the same weight p on the one hand, and only one of these two individuals with the weight $2p$.

Usually, the weights p_i are organised on the diagonal of a matrix denoted D of size (I, I), with the extra-diagonal terms being null. Matrix D is diagonal, which explains the origin of the term *diagonal metric*.

The metric defined in this way is Euclidean (that is, it is associated with a scalar product). By denoting $\langle u, v\rangle_D$ the scalar product in $\mathbb{R}^I$ between vectors v and w (the letter D refers to the use of weights p_i), we therefore obtain:

$$\left\langle \overrightarrow{OM_k}, \overrightarrow{OM_j} \right\rangle_D = \sum_i p_i(x_{ik} - \bar{x}_k)(x_{ij} - \bar{x}_j) = \mathrm{Covariance}[k, j].$$

This clearly shows that it is the centring which makes it possible to interpret this scalar product as a covariance. If, in addition, the variables are reduced, this scalar product is therefore equal to the cosine of the angle θ_{kj} between $\overrightarrow{OM_k}$ and $\overrightarrow{OM_j}$ and is interpreted as the correlation coefficient.

The above relationships are expressed as follows, denoting v' the transpose vector of v and v_k the kth column of X:

$$d^2(O, M_k) = ||v_k||_D^2 = v_k' D v_k.$$

The total inertia of cloud N_K in $\mathbb{R}^I$, with respect to the origin O, is easy to calculate as the variables all have an equal weight of 1.

$$\mathrm{Inertia}(N_K/O) = \sum_k 1\, d^2(O, M_k) = \sum_k \mathrm{Var}[k].$$

This total inertia is equal to those of cloud N_I in $\mathbb{R}^K$: the number K of variables in the centred-reduced case. This highly important property contributes to the duality between the table's rows and columns. Analysing the diversity of the individuals' profiles (cloud N_I) or the correlations between variables (cloud N_K) means examining clouds with the same inertia, a property induced by the fact that the points of one are the same as the dimensions of the space within which the other evolves.

1.4 Centring and Reducing

In PCA, the data table is always centred, as denoted by

$$x_{ik} \leftarrow x_{ik} - \bar{x}_k,$$

where x_{ik} designates the general term of the analysed table.

Within space $\mathbb{R}^K$, centring is geometrically interpreted as the positioning of the origin of the axes to the mean point G_I: this operation does not alter the shape of the studied cloud N_I. Within the space $\mathbb{R}^I$, centring is interpreted as the projection of N_K on the subspace orthogonal to the first bisector (the bisector is the line containing the constant functions).

Let us outline the proof of this second result, which is widely used in statistics. Let $\mathbf{1}$ be the vector of $\mathbb{R}^I$ (belonging to the first bisector), all the components of which have a value of 1. With the metric D, this is a unit vector. Let $\mathbf{1}^{\perp}$ be the subspace of $\mathbb{R}^I$, supplementary orthogonal of $\mathbf{1}$. Vector v (mean $\bar{v}$) of $\mathbb{R}^I$ can be decomposed into its projection on $\mathbf{1}$ (denoted $P_1(v)$) and on $\mathbf{1}^{\perp}$ (denoted $P_{1^{\perp}}(v)$). Thus

$$P_{1^{\perp}}(v) = v - P_1(v).$$

It is easy to show that $P_1(v)$ is the constant vector for which each component has a value of $\bar{v}$. The value to the right of the = sign is therefore the centred variable v. Thus, in PCA, cloud N_K evolves within $\mathbf{1}^{\perp}$, a subspace of dimension $I - 1$.

When the variables are not expressed in the same units, the data must be reduced as follows,

$$x_{ik} \leftarrow \frac{x_{ik} - \bar{x}_k}{s_k}.$$

The PCA is then said to be *standardised* (if the data are only centred, the PCA is said to be *unstandardised*). When the variables are expressed in the same units, the opportunity to reduce must be discussed on a case-by-case basis. In practice, unless there is a specific reason, users tend to reduce the data because, as we show, this balances the influence of each variable.

Within the space of individuals, reduction is interpreted geometrically as taking the standard deviation s_k as a unit of measurement for variable k. Within the variables' space, this means representing variable k by the unit vector for the direction linking O to M_k. The cloud N_K is therefore situated on a hypersphere with a radius of 1 (see Figure 1.3B).

1.5 Fitting Clouds N_I and N_K

If we could visualise clouds N_I and N_K perfectly, as is possible in the case of two dimensions, we would be able to answer most of our questions; examining N_I would show the multidimensional variability of the individuals, and N_K

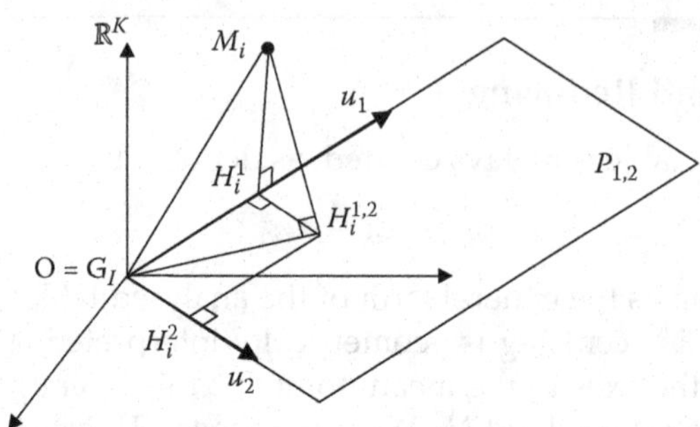

FIGURE 1.4
Fitting of individual i in $\mathbb{R}^K$.

would show the correlations between all of the variables taken two by two. But beyond three dimensions, the shape of the cloud becomes inaccessible to our senses. The aim of factorial methods in general, here of PCA, is to give, in a low-dimensional space, an approximate image of a cloud of points evolving within a high-dimensional space. This is often referred to as *dimension reduction*. We instead use the term *fitting*, which is widely used in statistics.

1.5.1 General Principles and Formalising Criteria

In factorial analysis, fitting a cloud means projecting it onto a set of orthogonal axes of maximum inertia.

Fitting N_I in $\mathbb{R}^K$

By denoting u_s a unit vector of the axis of rank s (in $\mathbb{R}^K$) and H_i^s the projection of point M_i on u_s, the criterion satisfied by u_s is

$$\sum_i p_i \, (OH_i^s)^2 \text{ maximum},$$

with the constraint of being orthogonal to $s-1$ directions already found, thus

$$u_s \perp u_t \quad \text{for} \quad t < s.$$

We therefore first look for u_1, the direction of maximum inertia, then u_2, the direction of maximum inertia orthogonal to u_1, and so on. The u_s axes are taken two by two to make up the factorial planes, particularly the first one (u_1, u_2). Figure 1.4 illustrates this fitting. H_i^1 (H_i^2, respectively) is the projection of M_i on u_1 (u_2, respectively), axis of rank 1 (rank 2, respectively). By combining the two coordinates of these projections, we obtain the projection, denoted $H_i^{1,2}$, of M_i on the plane $P_{1,2}$ generated by u_1 and u_2.

Matrix Formalisation. We denote X the data table with dimensions (I, K). $||OH_i^s||$ is obtained by the scalar product between u_s and the data vector i, that is to say the ith row of X. These projections (or more precisely their coordinates) are organised in vector F_s, of dimension I. Thus

$$F_s = Xu_s.$$

Expressing it in this way highlights the fact that F_s is a linear combination of initial variables, in which vector u_s contains the coefficients of this combination.

With the weights p_i (organised in diagonal matrix D), the criterion (to be maximised) is expressed:

$$F_s' D F_s = u_s' X' D X u_s.$$

In $X'DX$, we recognise the correlation matrix when the variables are standardised, and the covariance matrix when they are simply centred.

Fitting N_K in $\mathbb{R}^I$
By denoting v_s a unit vector on the axis of rank s (in $\mathbb{R}^I$), and H_k^s the projection of point M_k on v_s, the criterion satisfied by v_s is:

$$\sum_k (OH_s^k)^2 \text{ maximum},$$

$$\text{with } v_s \perp v_t \quad \text{for } t < s.$$

Matrix Formalisation. The coordinate of projection H_k^s is obtained by the scalar product between v_s and the column-vector of X containing the data of variable k (vector here denoted v_k)[1]. Accounting for metric D, we obtain:

$$OH_k^s = \langle v_k, v_s \rangle_D = v_k' D v_s.$$

The coordinates of the projections H_k^s are organised in vector G_s of dimension K. Thus

$$G_s = X' D v_s.$$

The criterion is expressed:

$$G_s' G_s = v_s' D X X' D v_s.$$

In XX', we recognise the matrix of scalar products between individuals.

1.5.2 Interpreting Criteria

In $\mathbb{R}^K$, due to the centring, the origin is at the mean point of cloud N_I. The criterion is then interpreted as the variance of projections. Within this space, we are therefore looking for the dimensions of maximum variance (or variability). This corresponds perfectly to the initial objective of describing the individuals' variability.

[1]For simplicity's sake, we use the same letter, v, to designate a standardised principal component (v_s or v_t) or an initial variable (v_k). They are indeed vectors of the same space and the indices, as well as the context, remove any ambiguity.

As the vectors u_s are orthogonal, the projection variances can be accumulated from one axis to another. If we add together these variances for all the axes, we obtain the total inertia of cloud N_I. The issue we initially put forward was to study the variability of individuals, or in other words, this total inertia. Unlike analyses conducted variable by variable, this approach is multidimensional insomuch as it decomposes this same total inertia by favouring the dimensions of $\mathbb{R}^K$ (that is, linear combinations of variables) which express the majority of this variability.

In $\mathbb{R}^I$, the origin of the axes is not located at the centre of gravity of N_K: in this space, the criterion is not interpreted in the same way as in $\mathbb{R}^K$. When the variables are centred-reduced, OH_k^s is the cosine of the angle between v_s and $\overrightarrow{OM_k}$, and is therefore interpreted as the correlation coefficient (denoted $r(k, v_s)$) between variable k and v_s. The criterion to be maximised is therefore expressed:

$$\sum_k [r(k, v_s)]^2.$$

This criterion can be interpreted as an indicator of the relationship between the function (on I) v_s, and on the other hand the set of K initial variables (it is worth 0 if v_s is uncorrelated with each of the K variables). It expresses that v_1 is the function on I (we show that this function is a linear combination of the initial variables) which is the most closely related (in terms of this criterion) to the initial variables. In the same way, v_2 is the function on I, uncorrelated to v_1, which is the most closely related to K and so on. This does indeed correspond to the initial objective of searching for synthetic variables.

When the variables are not reduced, OH_k^s is the covariance between k and v_s, and the criterion can be expressed:

$$\sum_k \mathrm{Var}[k][r(k, v_s)]^2.$$

Thus, an unstandardised PCA can be considered as a PCA on centred-reduced data in which each variable is attributed the same weight as its variance. This point of view is valuable when deciding whether to reduce the data.

1.5.3 Solution

In the Individuals' Space

In $\mathbb{R}^K$, we are looking for u_s which maximises the quantity

$$F_s' D F_s = u_s' X' D X u_s,$$

with the following norm and orthogonality constraints:

$$||u_s||^2 = u_s' u_s = 1 \quad \text{and} \quad \langle u_s, u_t \rangle_D = u_s' u_t = 0 \quad \text{for } t < s.$$

It can be shown that the vector u_s for which we searched verifies

$$X'DXu_s = \lambda_s u_s \quad \text{with} \quad \lambda_s = F_s' DF_s = u_s' X' DXu_s.$$

Thus, u_s is the unit eigenvector associated with the eigenvalue λ_s of $X'DX$, as the eigenvalues are ranked in descending order (they are equal to the inertia that we want to maximise). As usual, by *matrix diagonalisation*, we designate the procedure which calculates all of a matrix's eigenvalues and eigenvectors. The PCA is therefore based on diagonalising the correlation matrix in standardised cases and the covariance matrix if not.

Once the vector u_s has been obtained, the coordinates of the projection of individuals on the axis of rank s are obtained by

$$F_s = Xu_s.$$

Vector F_s, which contains the coordinates of the individuals on the axis of rank s, is known as the principal component of rank s (or the factor on I of rank s). Because it is a linear combination of (centred) initial variables, F_s is centred. It is easy to demonstrate that the variance of F_s is equal to λ_s.

Finally, we obtain

$$\sum_s \lambda_s = \text{trace}(X'DX) = \sum_k \text{Var}[k].$$

The idea of decomposing the total inertia of N_I into privileged dimensions can be clearly seen here.

In the Variables' Space

In $\mathbb{R}^I$ endowed with the metric D, we are looking for v_s which maximises

$$v_s' DXX' Dv_s,$$

with the following norm and orthogonality constraints

$$\begin{aligned} ||v_s||_D^2 &= v_s' Dv_s = 1, \\ \langle v_s, v_t \rangle_D &= v_s' Dv_t = 0 \quad \text{for } t < s. \end{aligned}$$

It can be shown that the vector v_s for which we searched verifies:

$$XX' Dv_s = \lambda_s v_s \quad \text{with } \lambda_s = v_s' DXX' Dv_s.$$

Thus, v_s is the unit eigenvector associated with the eigenvalue λ_s of $XX'D$, as the eigenvalues are ranked in descending order (they are equal to the inertia we want to maximise). It should be noted that v_s is a linear combination of initial variables (in Section 1.5.4 we show that its coefficients are in $X'Dv_s$). Here we are close to the perspective of $\mathbb{R}^K$ previously mentioned about F_s (as a combination of variables). The relationship between F_s and v_s is formalised in Section 1.5.4.

Once the vector v_s has been obtained, the coordinates of the projection of variables on the axis of rank s are obtained by

$$G_s = X'Dv_s.$$

Vector G_s, which contains the coordinates of the K variables on the axis of rank s, is known as the factor on K of rank s.

1.5.4 Relationships Between the Analyses of the Two Clouds

From the equation which yields the solution of the fitting in $\mathbb{R}^K$:

$$X'DXu_s = \lambda_s u_s,$$

we deduce:

$$XX'DXu_s = \lambda_s Xu_s$$

or

$$XX'DF_s = \lambda_s F_s.$$

This illustrates two essential results.

1. λ_s, here defined as the eigenvalue of $X'DX$, is also the eigenvalue of $XX'D$, therefore justifying the use of the same notation for the eigenvalues from analyses of N_I and N_K. Thus, the projected inertia of N_I on u_s (in $\mathbb{R}^K$) is equal to the projected inertia of N_K on v_s (in $\mathbb{R}^I$). We already saw that these two clouds have the same total inertia, a property which we classify under the term *duality*. Here, duality is considerably enriched.
2. If we retain all the eigenvalues, diagonalising $XX'D$ yields a perfect representation of the cloud of individuals (in its principal axes rather than in the basis of initial variables). This matrix therefore contains all the information, in terms of the shape of the cloud of individuals, and can thus represent it. This property is used (in MFA; see Chapter 7) to compare the clouds of points representing the same individuals in different spaces; indeed, the matrices $XX'D$ of the different clouds have the same dimensions and are thus comparable with one another.

Vector F_s, like v_s, is an eigenvector of $XX'D$ associated with the eigenvalue of rank s. The difference between the two is that v_s is standardised (v_s is said to be the *standardised principal component*) thus

$$v_s = \frac{1}{\sqrt{\lambda_s}} F_s = \frac{1}{\sqrt{\lambda_s}} Xu_s.$$

This relationship shows that the direction u_1 of $\mathbb{R}^K$ which best expresses the variability of N_I, corresponds to the 'best' synthetic variable v_1 (element of $\mathbb{R}^I$). Here, in the results of the PCA, we can see the intrinsic duality of the objectives. If a direction (u_s) of $\mathbb{R}^K$ expresses a great deal of inertia, the distribution of the individuals in this direction (F_s) is similar to that of many of the variables ($r^2(F_s, k)$ is high for many k) and F_s can be considered a synthetic variable.

We can continue this reasoning by switching the roles played by the rows and the columns. In order to do this, we bring together the coordinates of the variables (projected) on v_s in vector G_s (of dimension K) thus (see the end of Section 1.5.3)

$$G_s = X'Dv_s.$$

By expressing v_s in terms of u_s and by using the fact that u_s is the eigenvector of $X'DX$ associated with the eigenvalue λ_s, we obtain:

$$G_s = \frac{1}{\sqrt{\lambda_s}} X'DXu_s = \sqrt{\lambda_s} u_s.$$

This relationship shows that, up to a coefficient, the coordinates of the K variables on v_s (in $\mathbb{R}^K$) are the coefficients of the linear combination of variables defining u_s (in $\mathbb{R}^I$). This relationship is vital in the interpretation of axes. Indeed, we can consider two possible ways of interpreting linear combinations of variables:

1. The coefficients which define the combination
2. The initial variables to which this combination is linked

The previous relationship shows that these two approaches lead to the same result.

The relationships linking F_s and v_s on the one hand and G_s and u_s, belong to the aforementioned duality relationships (and indeed are the most remarkable elements of this duality): the projection of N_I on the one hand and N_K are the two sides of the same analysis. They can be summarised by saying that the factorial axes of one space are the factors of the other.

We can also link factors together. By expressing G_s in terms of F_s, we obtain

$$G_s = \frac{1}{\sqrt{\lambda_s}} X'DF_s.$$

For the kth coordinate, this equation is written

$$G_s(k) = \frac{1}{\sqrt{\lambda_s}} \sum_i p_i x_{ik} F_s(i).$$

This result was already mentioned: in the case of standardised PCA, the coordinate of variable k (on v_s) is the correlation coefficient between itself and F_s.

By expressing F_s in terms of G_s, we obtain

$$F_s = \frac{1}{\sqrt{\lambda_s}} XG_s.$$

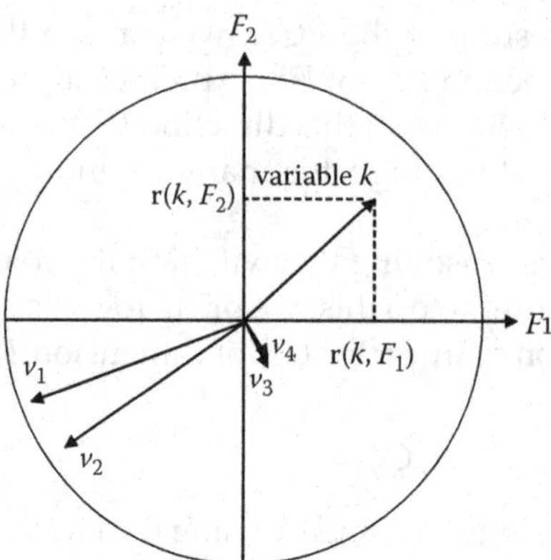

FIGURE 1.5
Correlation circle. r: correlation coefficient.

For the kth row, this equation is written

$$F_s(i) = \frac{1}{\sqrt{\lambda_s}} \sum_k x_{ik} G_s(k).$$

Along the axis of rank s, individuals have higher coordinates when they have high values for variables which are positively correlated to F_s **and** low values for variables which are negatively correlated to F_s (the *and* is **shown in** as users often ignore this second part). This property is commonly used (though often implicitly) in interpreting factorial planes. In this format, these relationships of duality are known as *transition relations* (understood from one space to the other), a term which was historically introduced within the context of correspondence analysis.

1.5.5 Representing the Variables

In standardised PCA, the 'universally' accepted representation is that of correlation circles, in which the coordinate of variable k on the axis of rank s is the correlation coefficient between this variable and the principal component F_s (see Figure 1.5). In standardised PCA, this representation coincides with the projection of cloud N_K.

This representation can be used to identify quickly which variables are the most closely correlated with each axis, either positively or negatively. When a variable presents its two coordinates scarcely different from 0 (that is, when it is close to the origin), it is orthogonal to the factorial plane. Additional information on this representation is given in Section 1.6.3.

In unstandardised PCA, the correlation circle and the projection of N_K do not coincide because, in the second case, the projection of a variable on an axis is interpreted as covariance. In such cases, analysis is therefore (slightly) more complicated.

Finally, both in standardised and unstandardised PCA, we might consider representing variable k on axis s by its coefficient in the linear combination which defines axis s, that is to say the kth component of u_s. The previous section shows that, axis by axis, this representation is proportional to the projection of N_K with coefficient $1/\sqrt{\lambda_s}$.

However, the kth component of u_s is equal to the projection (on u_s) of the unit vector representing variable k in $\mathbb{R}^K$. This is why some users superimpose this representation on that of the individuals, thus supplemented with a projection of the base vectors (of $\mathbb{R}^K$) on the same subspace. In practice, this representation has the major drawback of not being able to incorporate supplementary variables.

1.5.6 Number of Axes

In $\mathbb{R}^K$, there is a maximum of K orthogonal axes. Furthermore, to represent a set of I points perfectly, at most $I - 1$ axes (the origin is in G_I) are needed. The maximum number of axes (with an inertia of positive value) is therefore $\min\{I - 1, K\}$.

In $\mathbb{R}^I$, as the variables are centred, they are in a space of $I - 1$ dimension (in which there is a maximum of $I - 1$ orthogonal axes). Furthermore, the K variables generate a subspace with at most K dimensions. Again here, the maximum number of axes of inertia with a positive value is therefore $\min\{I - 1, K\}$.

1.5.7 Vocabulary: Axes and Factors

The term *axis* is not ambiguous: it represents a direction in a space. But the word *factor* is used with different meanings in different fields; for example, in R software, it refers to a qualitative variable.

This book uses the concept of factor as introduced by J.-P. Benzécri. When projecting a set of points on an axis, the coordinates of the projected points are brought together in a vector which we refer to as a factor. Thus, in $\mathbb{R}^K$ space, we project the set I of individuals on the axis of rank s and the coordinates of the projections are brought together in the vector denoted F_s; F_s is known as the factor on I of rank s. In space $\mathbb{R}^I$, the projection on axis s of all the K variables leads to vector G_s which is therefore the factor on K of rank s. Due to transition relations, the factor F_s (and G_s, respectively) calculated in $\mathbb{R}^K$ (and $\mathbb{R}^I$, respectively) is collinear with the axis of rank s in $\mathbb{R}^I$ (and $\mathbb{R}^K$, respectively) on which the variables (and individuals, respectively) are projected. In summary, we can say that, in factorial analysis, the factors in one space are the axes in the other.

In PCA, F_s is known as the principal component. The squared norm of F_s is equal to the sth eigenvalue. When F_s is standardised, it is called the *standardised principal component*: it is the vector v_s in $\mathbb{R}^I$ on which variables are projected. Often, we extend the name *principal component* to all the factors on I, regardless of the method (MCA, FAMD, etc.) that produced it.

1.6 Interpretation Aids

1.6.1 Percentage of Inertia Associated with an Axis

By dividing the projected inertia of cloud N_I (or N_K) onto the axis of rank s (λ_s) by the total inertia (equal to K in standardised PCA), we obtain the percentage of inertia associated with an axis; this percentage is used to measure:

- The quality of representation of clouds (N_I or N_K) by the axis of rank s
- The relative importance of axis s (compared to the other axes)

The percentage of inertia does not measure the advantage of a given axis for the user, firstly because it must be compared to the possible number of axes with an inertia of positive value. Thus, for example, 70% will not be considered in the same way if it is the result of the analysis of a table with the dimensions (5, 4) or (50, 40). In concrete terms, in standardised PCA, the percentage of inertia of an axis can be compared to what it would be in the case of absence of structure in the data (spherical cloud of individuals, with no particular direction of extension, or variables uncorrelated two by two, which amounts to the same thing) or $100/K$ which corresponds to an eigenvalue of 1. Consequently, we must approach with caution when faced with an axis associated with an eigenvalue of less than 1, which represents less variability than one single initial variable.

In the same context, we can simulate tables (with fixed dimensions) from independent variables. We thus obtain a distribution of the first eigenvalue in which we can situate a first observed eigenvalue. To do this, we apply a common (unilateral) test approach, with the following H_0 hypothesis: the data stem from a process in which the variables are independent. Concretely, tables were drawn up to give the 95% quantile of the distributions of the first eigenvalue for different values of I and K. Even if, in a given case, this hypothesis does not have a clear meaning, the value of such a table is always useful, at least for information.

The percentages of inertia can be added over several axes; by adding these percentages for the first two axes for example, we measure:

- The representation quality of clouds (N_I or N_K) by the first plane
- The relative importance of the first plane (compared with other planes, or with axes considered individually)

1.6.2 Contribution of One Point to the Inertia of an Axis

Case of an Individual

In the maximised quantity (projected inertia of N_I), we can individualise the role of each individual, known as their contribution. This contribution is generally expressed as a percentage (of total inertia) thus, for individual i and the axis of rank s,

$$\text{Contribution}(i, s) = \frac{\text{Projected inertia of point } i \text{ on } s}{\text{Projected inertia of } N_I \text{ on } s} = \frac{p_i (OH_i^s)^2}{\lambda_s}.$$

The notion of contribution is important in order to identify specific situations in which one axis arises from very few individuals, or one single individual. That said, with that in mind, when individuals are of the same weight, it is sufficient to look at the individuals on the factorial plane. Measuring contribution is therefore only truly useful when the weights of individuals are different.

Case of a Variable

In centred-reduced cases, the 'raw' contribution (that is, not expressed as a percentage of total inertia) of variable k on the inertia of axis of rank s is equal to the square of its correlation coefficient with the principal component F_s and is therefore read directly on the correlation circle. This is not the case in unstandardised PCA as the raw contribution is then a covariance. In this case, the two types of graphs (projection of N_K and correlation circle) are required for interpretation. This is one of the reasons why unstandardised PCA is considered to be (slightly) more complicated to interpret than standardised PCA.

Comment. For a given axis, the contributions can be added together to represent the contribution of a subset of individuals or a subset of variables.

1.6.3 Quality of Representation of a Point by an Axis

A point (individual i or variable k) can be attributed the percentage of inertia introduced in Section 1.6.1 for a cloud. We thus measure the quality of representation (of the inertia) of a point by an axis. Thus, for an individual i and axis of rank s (see Figure 1.4):

$$\text{Qlt}(i, s) = \frac{\text{Projected inertia of } i \text{ on } u_s}{\text{Total inertia of } i} = \frac{(OH_i^s)^2}{(OM_i)^2} = \cos^2(\overrightarrow{OM_i}, u_s).$$

For one point, this indicator can be added over a number of axes (in the same way as inertia percentages), making it possible to measure the representation quality of a point by a plane, for example.

In the case of variable k (whether the PCA is standardised or not), this indicator can be confused with the squared correlation coefficient between

k and the principal component of rank s (already denoted $r(k, F_s)^2$). In the *correlation circle* representation, the quality of representation of a variable by the plane is evaluated visually by the distance between the point representing the variable (generally the end of an arrow) and the correlation circle. It is therefore not necessary to draw up indicator tables for the variables' quality of representation.

Thus, in Figure 1.5, variables v_1 and v_2 are well represented and the angle which they form gives a good idea of the correlation coefficient (close to 1). However, variables v_3 and v_4 are poorly represented and we can infer nothing from their proximity on the graph.

In practice, for the individuals, this indicator is mainly used to select a few individuals with the aim of illustrating an axis: if an individual is well represented by the axis, its 'particularity' (that is to say its deviation from the mean point) mainly refers to the axis and it will be easy to link its coordinate (on this axis) with its data.

1.7 First Example: 909 Baccalaureate Candidates

Here we comment on the results of the standardised PCA conducted on the aforementioned table of the five grades received on the Baccalaureate by 909 students. Table 1.6 given at the end of the chapter brings together the data for a number of individuals mentioned in the text and on the graphs. Compared to the data files, this table is transposed in order to write the complete labels of the variables.

1.7.1 Projected Inertia (Eigenvalues)

With 5 variables and 909 individuals, there are a maximum of 5 factorial axes of nonzero inertia (see Section 1.5.6). The decrease in eigenvalues (see Table 1.1 along with Figure 1.6) reveals a predominant factor. Moreover, only the first eigenvalue has a value of more than 1. For this reason, we might think that only the first axis should be retained for interpretation. In fact, we retain more:

TABLE 1.1
Bacc. PCA. Eigenvalues and Percentages of Inertia

Axis	Eigenvalue	Percentage of Inertia	Cumulative Percentage
1	2.4081	48.16	48.16
2	0.9130	18.26	66.42
3	0.6623	13.25	79.67
4	0.6419	12.84	92.51
5	0.3747	7.49	100.00

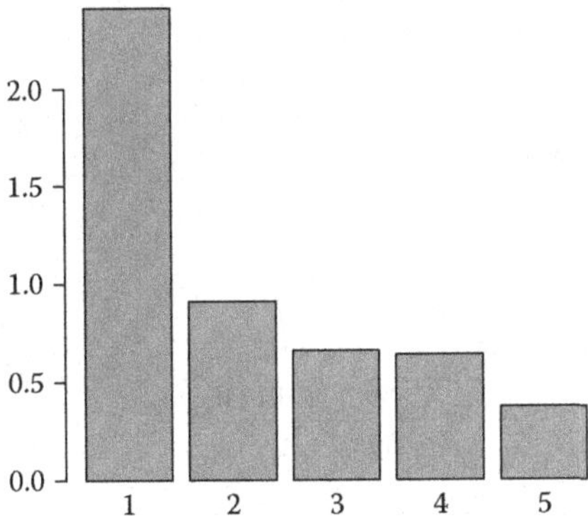

FIGURE 1.6
Bacc. PCA. Barplot of eigenvalues.

this is an interesting feature of this example, which is food for thought on how to select the axes to retain for interpretation.

1.7.2 Interpreting the Axes

Having checked the regular shape of the cloud of individuals on the factorial planes (in other words, that no axis comes from a small number of individuals alone, in which case the interpretation would be conducted first in terms of the individuals), we can make interpretations using the correlation circles (see Figure 1.7).

The first axis is a *size effect*: all the variables are positively correlated with one another and thus with the first principal component; this axis opposes the students who obtained high grades in all subjects (such as 264) and those who obtained low grades in all subjects (such as 863). This axis can be interpreted as the *general level* (of the student), or the *general average*. Indeed, if we calculate the general average (without coefficients) of the five subjects, we observe a correlation coefficient of .9992 between this average and the first principal component, thus validating the interpretation.

The second axis opposes scientific subjects (*maths* and *physics*) with literary subjects (*philosophy* and *history–geography*), and at the same time, the students with a scientific profile, that is, subjects having higher grades in the scientific subjects than in literary subjects (such as 850), with the students with a literary profile (such as 46).

The simplest way to reach this interpretation is to consider the coordinate of a variable (on the axis of rank s) as its coefficient (up to the coefficient $\sqrt{\lambda_s}$) in the linear combination defining u_s (see Section 1.5.4); here, for axis 2: $u_2 = 0.5$ maths $+ 0.42$ physics – 0.50 history–geography – 0.47 philosophy – 0.14 natural sciences, which can be assimilated with:

$$\tilde{u}_2 = \frac{1}{2}(\text{maths} + \text{physics}) - \frac{1}{2}(\text{hist–geo} + \text{philosophy})$$

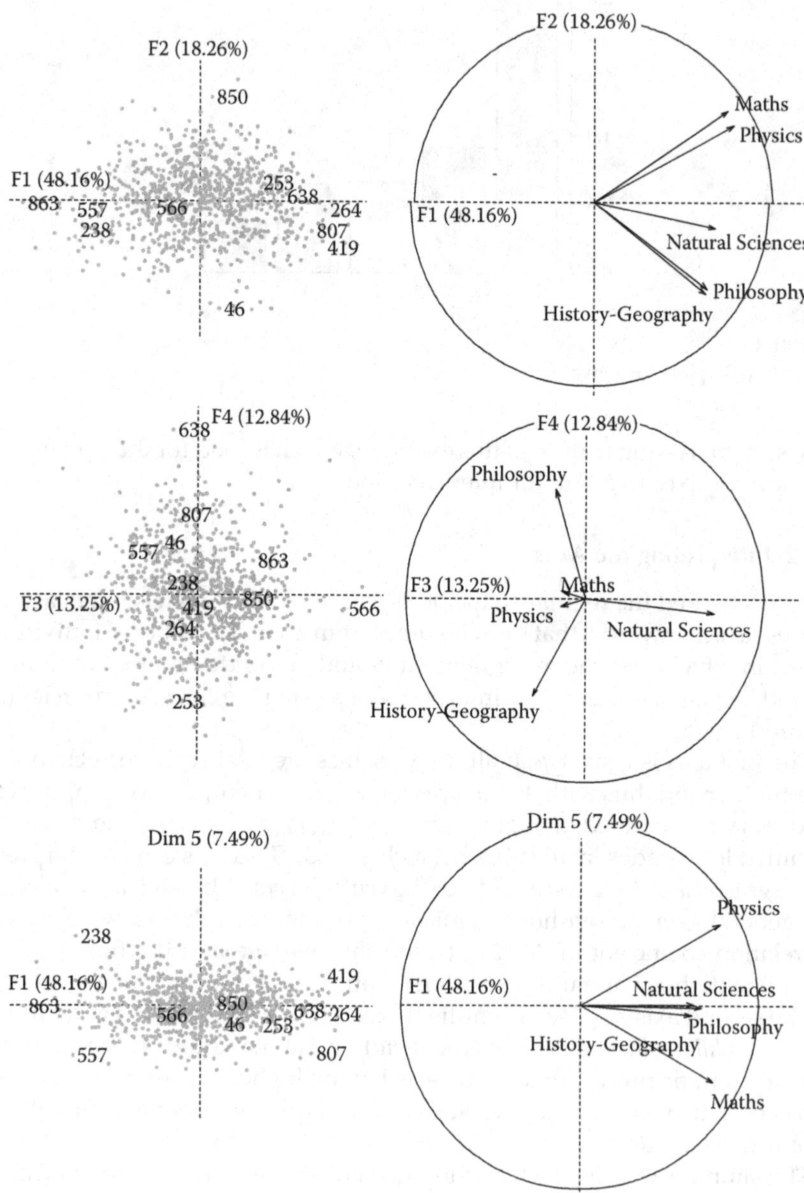

FIGURE 1.7
Bacc. PCA. Representation of individuals and variables on three factorial planes; a few individuals are represented by their ordinal number in the file.

This clearly shows that the students with a positive coordinate for the second axis do not have a high grade (in itself) in scientific subjects but have a higher grade in scientific subjects than in literary subjects (for example, compare 253 and 264: the scientific profile of 253 can in part be attributed to the low grade in *philosophy*).

If we calculate the linear combination of the variables defined by $\tilde{u}_2$, we obtain a new variable whose correlation coefficient with the second principal component (F_2) is worth 0.963, thus validating the interpretation.

The third principal component is essentially correlated with the grade in natural sciences, which is the subject the least well represented on the first plane. We could therefore name this axis *Specificity of natural sciences*. It is important to differentiate between this axis and the *natural sciences* variable itself, as here we must reason 'orthogonally to axes 1 and 2.' In other words, this axis opposes the students (such as 566) who obtained a high grade in *natural sciences* (high compared to their other grades) with those (such as 557) who scored low in this subject (low compared to their other grades). Again, we consider the linear combination corresponding to the third vector (in $\mathbb{R}^K$) u_3:

u_3 = 0.72 natural sciences − (0.13 maths + 0.14 physics + 0.29 history–geography + 0.16 philosophy),

which we could assimilate with:

0.7 [natural sciences - $\frac{1}{4}$(maths + physics + history–geography + philosophy)].

If we apply this last formula to the data, we obtain a new variable for which the correlation coefficient with F_3 is worth .9700, thus validating the interpretation.

The fourth axis confronts *philosophy* and *history–geography*. These two variables appear closely linked on the first plane, a relationship which participates both in the size effect and the notion of literary profile. Here, our position is at a constant general level and at a constant (scientific versus literary) profile, and we highlight profiles which are either *philosophy*-oriented or *history–geography*-oriented. If we calculate the *philosophy − history–geography* variable, we observe a correlation coefficient of 0.9864 between this variable and the fourth principal component (F4). As an example, to illustrate this fourth principal component, we could compare students 807 and 253.

The fifth axis confronts *maths* with *physics*, therefore the students who obtain a higher grade in *maths* than in *physics* (such as 238 or 419) with students with the opposite characteristic (such as 557 and 807). Applying the same reasoning as for the fourth axis to this fifth axis leads to the calculation of the *physics − maths* difference; the correlation coefficient between the variable resulting from this difference and the fifth principal component is worth .9884.

1.7.3 Methodological Remarks

Interpretability and Percentage of Inertia

Thus, in this analysis, all of the axes can be clearly interpreted. This is therefore an unusual case, which clearly highlights two important points:

1. PCA can be seen as a base change (to examine the variability of individuals beginning with the dimensions with the greatest inertia); in the most common applications, we use only the first axes of the new base, reducing the visibility of this point of view.
2. The interpretability of an axis is not necessarily associated with high inertia; here, the fifth axis is both clear (its interpretation is simple and the position of the individuals can be easily linked to the data) and marginal (it represents very little variability: the maths and physics grades are correlated ($r = .62$) and the deviation between the two corresponds to little variability).

Two Readings of the Correlation Circle

In the representation of variables, the coordinate of variable k along the axis of rank s can be considered in two different ways (see Section 1.5.4):

1. The correlation coefficient between variable k and principal component F_s
2. The coefficient of variable k in the linear combination defining axis u_s in $\mathbb{R}^K$ (up to the coefficient $\sqrt{\lambda_s}$)

Generally, the first approach is favoured. In this example, we mainly used the second, which is better suited to situations in which the linear combinations of variables can be interpreted easily. To do this, the variables must be expressed in the same units. It must be noted that the linear combinations presented above were calculated from raw variables, which yields a more striking interpretation. To better 'fit' the PCA, it is possible to apply these combinations to the centred-reduced variables, which does not change much in these data where the standard deviation varies only slightly from one variable to another (by doing this, the correlation coefficients between these combinations and the principal components increase slightly in four out of five cases).

In practice, a significant advantage of the first approach is that it can also be applied to (quantitative) supplementary variables (see the following section).

Validating the Interpretation

Analysing the representations of active variables led to interpretations in the form of combinations of initial variables. Then, in order to validate these interpretations, we calculate these combinations. This yields new variables which we introduce as supplementary elements. We describe this technique in detail in the next section. This practice is very general: the interpretation of an axis often suggests calculating combinations (which may or may not be linear) of variables, or introducing new variables, in order to validate the interpretation.

PCA and Synthetic Visualisation
PCA is mainly used to obtain a synthetic view of a data table; we 're-place' the initial variables K with two or three synthetic variables (the principal components). This point of view is called *dimension reduction.* However, in this example, it is tempting to retain the five axes as all five can be easily interpreted. As we start with only five variables, there is no dimension reduction and, from this 'countable' perspective, there is no synthesis.

Nevertheless, even if we retain all five axes, it is more interesting (in our opinion) to comment on the PCA than on the initial variables. We let you as the reader judge for yourself, but would like to draw your attention to two points. Unlike the initial variables, the principal components are:

1. Organised into a hierarchy (in descending order of variance)
2. Uncorrelated with one another

1.8 Supplementary Elements

An element, be it an individual or a variable, is said to be supplementary if it does not participate in constructing the axes on which it is represented. In practice, factorial analysis is almost always performed with supplementary elements, and in particular, supplementary variables. When faced with an available set of variables, the decision to consider a given variable as either active or supplementary is not always easy and the objectives of the analysis must be precisely specified. In real cases, the easiest approach is to reason in terms of individual profiles, and to ask oneself which variables participate in creating this profile.

Let us look back at the example of the 909 students. For these students we have access to:

- Baccalaureate grades for five subjects
- Five grades for these five subjects obtained during the school year, here called *year grades*
- The high school, a qualitative variable with 23 categories

The *high school* variable, being qualitative, cannot be active. In any case, we are studying school profiles and of course the high school is not part of that. Nonetheless, we do not believe this variable should be eliminated, as it is interesting to link the principal dimensions of variability of the school profiles with the high schools (see the following section).

There are three options for the other variables:

1. Define the school profiles from the Baccalaureate grades alone, for example, because they are more rigorously comparable from one student to another when we consider all high schools together, or simply because the Baccalaureate grades are the principal object of the study; retaining the year grades as supplementary means linking them to the principal dimensions of variability of the Baccalaureate profiles in order to enrich the interpretation (in concrete terms, in-depth interpretation of one dimension of variability of Baccalaureate grades will vary depending on whether it is closely correlated with the year grades, or not at all).
2. Define the school profile from the year profiles alone, for example, because the students' work during the year is the main object of the study; therefore retaining the Baccalaureate grades as supplementary means associating them with the principal dimensions of variability of the 'year' profiles.
3. Define the school profile from the set of 10 grades because we are not focussing on one subset.

In the example, the projection of year grades yields the graphs in Figure 1.8. On the first plane, the year grades closely follow Baccalaureate grades. This is not the case (or only very slightly) for the following axes. This suggests that the commentaries of the first axes refer to students' 'structural' profiles (a high grade is associated with a 'long-term' aptitude) and those of the following axes refer to 'temporary' profiles (a high grade corresponds to temporary success such as luck, cramming, etc.).

From this we retain:

- The need to specify the (active/supplementary) status chosen for each variable
- The advantage of the notion of individuals' profiles to make this choice; that is to say to connect it directly to the precise objective of the analysis
- The meaning of the supplementary status of a variable: link the variable to the principal dimensions of variability of the profiles

Remark

In $\mathbb{R}^I$, projecting a variable y on the plane defined by the first two axes (v_1 and v_2) means estimating the parameters of the multiple linear regression model expressing y in terms of v_1 and v_2. Indeed, as v_1 and v_2 are uncorrelated, their coefficients in the multiple regression are the same as those of the simple regression. And, when the (explanatory and response) variables are centred and reduced, the latter are equal to their correlation coefficient with y (which is why this coefficient is denoted r).

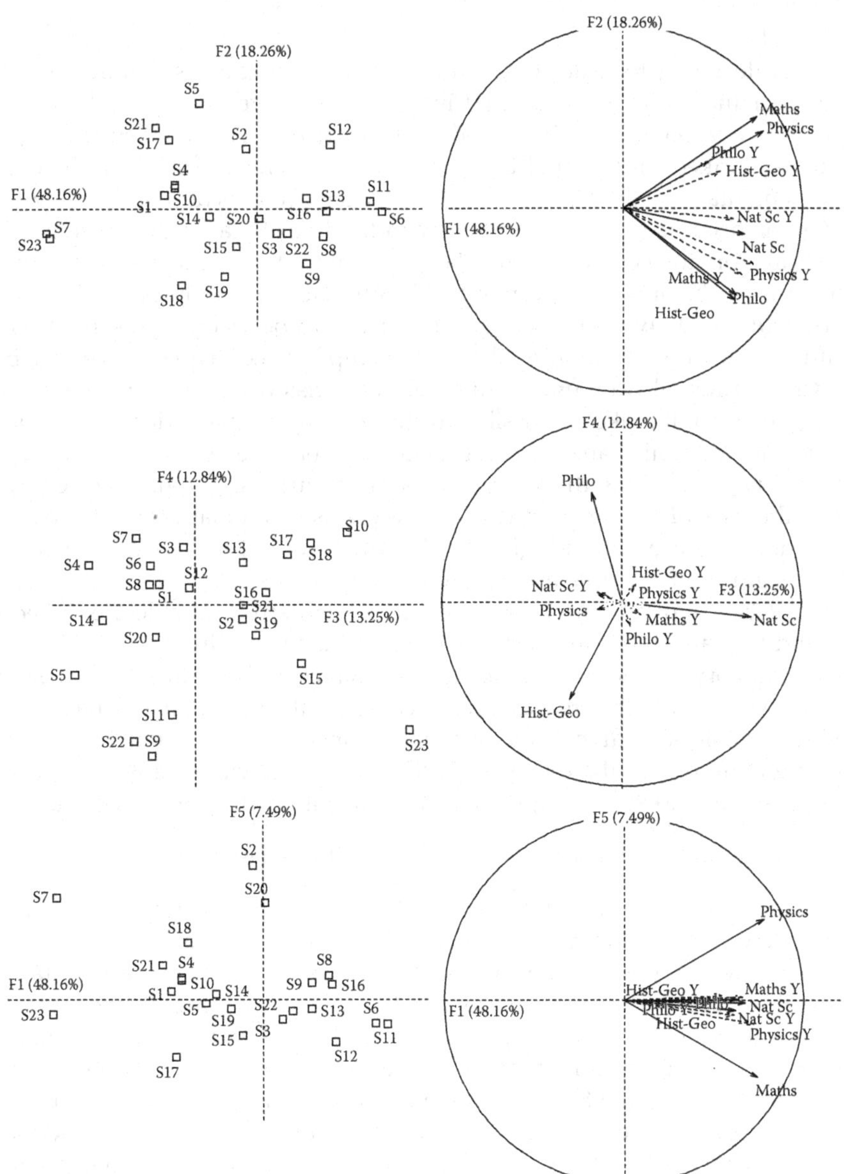

FIGURE 1.8
Bacc. PCA. Right: representations from Figure 1.7 to which we have added year grades as supplementary variables. Left: representation of high schools as their students' centres of gravity.

1.9 Qualitative Variables in PCA

Principle

PCA is designed to analyse several quantitative variables simultaneously. Thus, qualitative variables cannot intervene as active (elements). However, qualitative variables can be introduced as supplementary (elements), that is to say, linked to principal components. In concrete terms, introducing a supplementary qualitative variable in PCA can occur as described below.

On factorial planes, we can identify individuals by a symbol designating their category for a variable; in the Baccalaureate data, this means representing students by the numbers given to their high school, for example. This makes it possible to analyse precisely (visualising both between-high schools and within-high school variability) the relationship between a qualitative variable and the principal components (a level of precision which is indeed rarely necessary), but it makes it possible to study only a single variable at a time.

On the factorial planes, we can also represent the centres of gravity of individuals presenting the same category of a variable (for all the categories of all the variables); this procedure is less precise than that detailed above (it does not visualise within-high school variability) but it is used to visualise the categories of multiple qualitative variables at once.

Intuitively, a quantitative variable and a qualitative variable are linked if the individuals of the same class (a class gathers the individuals having the same category for the qualitative variable) have similar values for the quantitative variable. More precisely, we decompose the variability of the quantitative variable according to the partition defined by the qualitative variable. We therefore obtain the equation of the analysis of variance which, in our language, can be expressed (in which case we call it Huygens' theorem):

Total inertia = between-classes inertia + within-class inertia

The squared correlation ratio divides between-inertia by total-inertia. This is the percentage of inertia 'explained' by the qualitative variable.

Therefore, the overall intensity of the relationship between a qualitative variable q and a principal component F_s can be measured by the squared correlation ratio between q and F_s. We can then construct a representation of the variables using this measurement as a coordinate. This representation is particularly useful when there are a large number of qualitative variables, which is not the case in the example of Baccalaureate grades. It is described in more detail later and illustrated in multiple correspondence analysis (see Figure 2.3).

The significance of this relationship can be measured using the p-value associated with the overall F test of the factor effect in the one-way analysis of variance 'explaining' component F_s using variable q. As in a regular analysis of variance, this overall test is a prerequisite to detailed examination of the categories outlined below. When there are a high number of qualitative

TABLE 1.2
Bacc. PCA. Indicators of the Relationship Between the *High School* Variable, and the Principal Components and Initial Variables[a]

PCA (Factor)	p-Value	η^2	Variable Label	p-Value	η^2
F5	2.28E-21	0.156	Hist.-Geo. year	9.85E-28	0.187
F3	5.20E-15	0.123	Hist.-Geo. Bacc.	6.65E-18	0.138
F1	2.67E-12	0.108	Philo. year	2.71E-17	0.135
F2	5.60E-09	0.088	Maths Bacc.	3.89E-16	0.129
F4	2.58E-07	0.078	Bacc	5.02E-11	0.100
			Nat. Sciences Bacc.	1.06E-10	0.099
			Phys. year	1.90E-08	0.085
			Physics Bacc.	4.02E-08	0.083
			Philosophy Bacc.	7.41E-08	0.081
			Cont. Bacc.	1.08E-05	0.067
			Nat. Sciences year	1.82E-05	0.065
			Maths year	4.78E-05	0.062

[a] These indicators are calculated from the analysis of variance explaining a principal component, or an initial variable, from the high school. The p-value is that of F of the overall high school effect.

variables, it can be used to rank them in hierarchical order for each principal component.

The significance of the deviation, along axes of rank s, between (the centre of gravity of the individuals which carry) category j (of variable q) and the origin of the axes (the mean point G_I) can be evaluated in many ways. One helpful way is to use the p-value associated with the coefficient of category j in the analysis of variance mentioned above. An indicator like this is essential to accompany the representation of the centres of gravity of these categories as it takes into account, in addition to the deviation from the origin of the axes (the coordinate), the within-category variability as well as the frequency of categories (one same coordinate can be significant if it corresponds to a large frequency and not significant otherwise).

Baccalaureate Grades Example

The indicators between the (qualitative) high school variable and the principal components are brought together in Table 1.2.

All of the p-values for the PCA factors are highly significant: the *high school* variable is undoubtedly linked to all of the dimensions from the PCA. The percentages of variance explained by the high school may seem weak, but we must consider the context. For a given subject, we 'know' that, independently of the high school, there will be a great difference between the students' levels. The 'temporary' variability related to a specific exam is added to the 'structural' individual variability. The resulting individual variability is important. Where does the high school effect come from in this case? Such an effect may,

TABLE 1.3
Bacc. Statistics for Students from High School 2[a]

Grade	High School 2 Average	General Average	p-value
Maths Bacc.	11.737	13.207	0.117
Physics Bacc.	13.684	11.002	0.001
Maths year	10.456	10.950	0.493
Physics year	12.148	11.166	0.082

[a] Averages, in mathematics and physics, and p-value of the corresponding coefficients in the analysis of variance.

for example, be due to a selective entrance exam or students being more or less well prepared for the exam. When we take all of these factors into account, the observed percentages (between 7.8% and 15.6%) could be considered as being rather high.

Due to this set of significant relationships, we examine the relationships between the high schools and the initial variables. We highlight the same indicator values: overall, the high school explains 10% of the variability of each of the grades on the Bacc.

More precisely, each high school is represented on each factorial plane as its students' centre of gravity (see Figure 1.8). For example, according to its position on the plane (1,5), high school 2 (S2) is medium from the point of view of the overall set of results of its students, but very unusual from the point of view of the pair (mathematics, physics): its students scored much higher in physics than in mathematics, as verified and specified in Table 1.3.

Axis 5 presents a specific interest, as it is the last and is associated with an eigenvalue (0.38) of much less than 1; we have already shown (see Section 1.7.3) that its clear interpretation suggests it should be retained for comment despite its low inertia. This decision is backed up by its strong relationship with the *high school* variable.

Generally, a significant relationship between a principal component and an illustrative variable is a strong argument for not simply considering this component as noise. This is not a very strong argument in the particular case of the *high school* variable in this example, as this variable is linked to all of the initial variables. Nonetheless, it must be noted that the *high school* variable is most closely related to precisely this fifth principal component. Furthermore, the high schools which are the best characterised by this axis (2, 7, 20, 18) have only a small number of students (19, 11, 19, 20) which very probably correspond to one single class; a situation in which the *high school* effect could in fact be expressing a 'teacher' effect. These clues are given merely as an example; they illustrate how PCA in general, and particularly supplementary qualitative variables, can help in exploring data tables.

TABLE 1.4
The Six Orange Juices Studied

n°	Brand	Origin	Type
P1	Pampryl	Other	Ambient
P2	Tropicana	Florida	Ambient
P3	Fruvita	Florida	Refrigerated
P4	Joker	Other	Ambient
P5	Tropicana	Florida	Refrigerated
P6	Pampryl	Other	Refrigerated

1.10 Second Example: Six Orange Juices

The following data are the subject of many different analyses in this book. There are two objectives to the following example: to illustrate an interpretive approach using a small example, and to give a first glimpse of these data to appreciate better the more detailed analyses which are made in subsequent chapters.

Six pure orange juices (see Table 1.4) were chosen from the leading brands on the market (in 1997) which exist both as *refrigerated* (fr; in the store, these juices, which are less pasteurised, must be kept refrigerated) and *ambient* (amb; in the store, these juices are presented on normal shelves at room temperature). Three of these juices are made with Florida oranges (both Tropicana juices and the Fruvita juice).

These six juices were subjected to eight chemical measurements: two pH measurements, titre, citric acid, sugars and vitamin C. A total of 96 student engineers from a higher education establishment specialising in the food industry, who were used to tasting products and who were orange-juice drinkers, each described these six products according to seven descriptors: odour intensity, odour typicity, taste intensity, pulpy characteristic, sweetness, sourness and bitterness. They also expressed an overall hedonic evaluation. The data table (represented transposed in Table 1.5) confronts the six orange juices in the rows with, in the columns, the $8 + 7 + 1$ quantitative variables to which we add two qualitative variables, each with two categories: origin (Florida/other) and type (ambient/refrigerated).

Our first look at this focuses on the chemical variables. For a given product, its values for the eight measurements make up its *chemical profile*. To highlight the principal dimensions of variability for these chemical profiles, we perform a PCA in which the chemical measurements are introduced as active.

The sensory descriptors are introduced as supplementary with the aim of answering the question: are the principal dimensions of variability (of chemical data) related to sensory descriptors? Moreover, introducing the origin

TABLE 1.5
Orange Juice. Chemical and Sensory Data

	P1	P2	P3	P4	P5	P6	Average
Glucose (g/L)	25.32	17.33	23.65	32.42	22.70	27.16	24.76
Fructose (g/L)	27.36	20.00	25.65	34.54	25.32	29.48	27.06
Sucrose(g/L)	36.45	44.15	52.12	22.92	45.80	38.94	40.06
Raw pH	3.59	3.89	3.85	3.60	3.82	3.68	3.74
Refined pH	3.55	3.84	3.81	3.58	3.78	3.66	3.70
Titre	13.98	11.14	11.51	15.75	11.80	12.21	12.73
Citric acid	0.84	0.67	0.69	0.95	0.71	0.74	0.77
Vitamin C	43.44	32.70	37.00	36.60	39.50	27.00	36.04
Odour intensity	2.82	2.76	2.83	2.76	3.20	3.07	2.91
Odour typicity	2.53	2.82	2.88	2.59	3.02	2.73	2.76
Pulp	1.66	1.91	4.00	1.66	3.69	3.34	2.71
Taste intensity	3.46	3.23	3.45	3.37	3.12	3.54	3.36
Sourness	3.15	2.55	2.42	3.05	2.33	3.31	2.80
Bitterness	2.97	2.08	1.76	2.56	1.97	2.63	2.33
Sweetness	2.60	3.32	3.38	2.80	3.34	2.90	3.06
Overall evaluation	2.68	3.01	3.27	2.67	2.97	2.65	2.87

and the type of juice makes it possible to link these dimensions to these two variables. (Is the principal dimension of the chemical variability of the juices related to their origin? To their type?) The results of this PCA are presented in Figures 1.9 and 1.10.

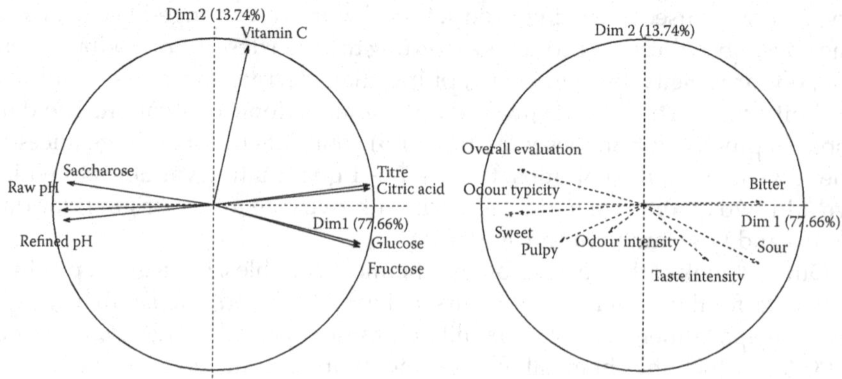

FIGURE 1.9
Orange Juice. PCA. Representation of active (left) and supplementary variables (right) on the first plane.

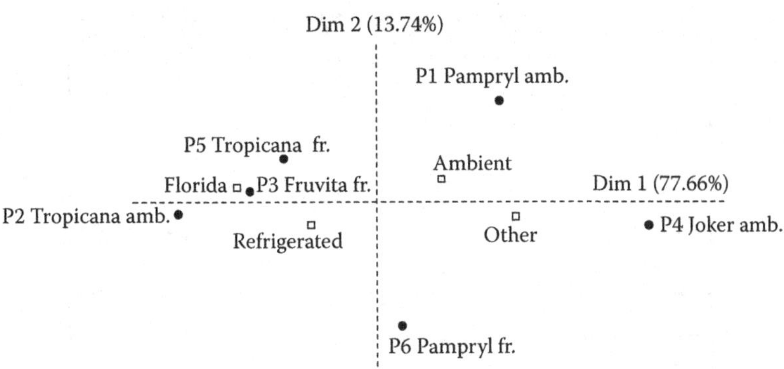

FIGURE 1.10
Orange Juice. PCA. Representation of the individuals and categories of the supplementary qualitative variables (in italics).

The first plane expresses 91.4% of the inertia, and we must limit ourselves to this.

The first axis is predominant (77.68%). It opposes the two pH measurements with the titre and the citric acid. Here this corresponds to juices 2, 3 and 5, with low acidity, in opposition with the other more acidic juices. The distribution of sugars is related to this acidity. The nonacidic juices are proportionally richer in sucrose. This opposition between sugars is related to the hydrolysis of sucrose (in glucose and fructose), hydrolysis which is accentuated in acidic environments. This axis can be summarised by 'acidity'. It is related to the juices' origins: the Florida juices are less acidic than the others. The second axis corresponds to vitamin C.

The representation of sensory variables shows a strong relationship between (measured) acidity and sensory description. The chemically acidic juices (4, 1, 6) are perceived as sour, but also bitter and not sweet. Conversely, the juices which are chemically not very acidic (2, 3, 5) are perceived as not sour but also not bitter, and sweet. Finally, the overall evaluation is closely related to the first factor: overall, the tasters preferred the sweet juices with low sourness and bitterness.

1.11 PCA in FactoMineR

The factorial methods described in this book are available in the R package **FactoMineR**. It is possible to use them directly, using lines of code, or by using **R Commander**. We begin by presenting the latter technique which is simple but, of course, less flexible. To illustrate our ideas we use the Baccalaureate data analysed previously in this chapter. These data were imported, either using the importation menu in **FactoMineR** or using the `read.table` function (see below).

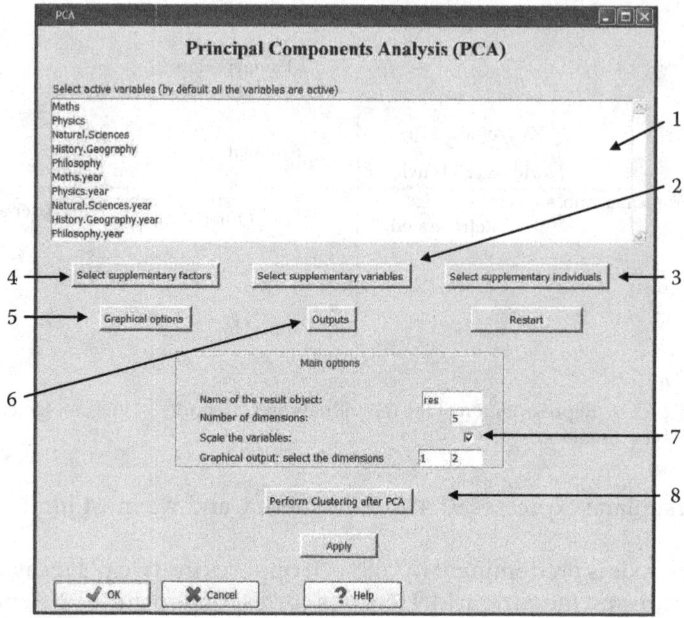

FIGURE 1.11
Main PCA menu in the drop-down **FactoMineR** menu.

Drop-Down Menu in R Commander

Main Menu (See Figure 1.11)

1. The (quantitative) active variables are selected in the main window. They are not necessarily adjacent. By default, all the (quantitative) variables are active.
2. By default there are no quantitative supplementary variables. By using the drop-down menu, we ignore the quantitative variables in the file which are not selected as active. This button opens a menu to choose the quantitative supplementary variables (in the same way as we select active variables in 1).
3. By default, all the individuals are active. This button opens a window containing the list of individuals in which the supplementary individuals are chosen.
4. In R terminology, the qualitative variables are called *factors*. By default there are no qualitative supplementary variables. This button opens a window containing the list of qualitative variables (a list here reduced to the `High School` variable). In this list, we choose the qualitative variables to introduce as supplementary. By using the drop-down menu, we ignore the qualitative variables in the file which are not selected.

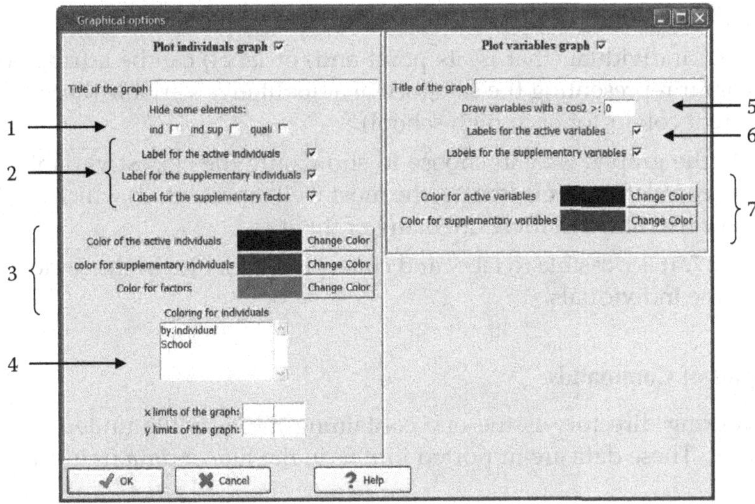

FIGURE 1.12
Graphical options window.

5. Open the window as seen in Figure 1.12.
6. Here we can select the results tables to be exported and specify a '.csv' file name (Excel-compatible file type) in which all of these tables will be brought together.
7. Reducing (here named `scale`) the variables means conducting a standardised PCA (the default option). By not reducing the data, we obtain a unstandardised PCA, in which each variable is attributed a weight equal to its variance.
8. Used to perform a clustering (Ward's method) from factorial coordinates, this sequence makes it possible to produce results using both those of the clustering and those of the PCA (such as a factorial plane in which the individuals are colour-coded according to which group they belong to by the partition defined from the hierarchical tree). This sequence of a classification following a factor analysis is available for all the factorial methods in **FactoMineR**. It is illustrated in MFA (Chapter 4).

Graphical Options (See Figure 1.12)

1. The elements represented on this type of graph can be the active and/or supplementary individuals and/or the barycentres of the individuals presenting one single category.
2. The elements can be labelled. In the example, the individuals are shown simply with a point whereas the high schools are labelled.

3. Different kinds of points can be attributed different colours.
4. Each individual (that is, its point and/or label) can be attributed a colour representing the category of a qualitative variable (here a different colour for each high school).
5. On the graphs we can choose to show only the closest variables to the correlation circle (that is, the most well represented) which can be very useful when there are many of them.
6. and 7. It is possible to label and colour the variables in the same way as the individuals.

Examples of Commands

The working directory is the one containing the data file under the name `Bac.csv`. These data are imported into R (in the file, or data.frame, `Bac`) by

```
> Bac=read.table("Bac.csv",header=TRUE,sep=";",dec=",",
+ row.names=1)
```

Edit the first two rows of the file to check the import.

```
> Bac[1:2,]

HighSchool MAT PHY NS HG PHI mat3t  phy3t  sn3t   hg3T   phi3T
1 L_4    11  12  10 7  5   12.50  9.83   12.60  10.63  9.70
2 L_4    15  12  10 11 13  13.47  11.13  10.83  11.33  10.23
```

The data are in the `Bac` data-frame. The eleven columns are in the following order: the high school, the five Baccalaureate grades, and the five year grades.

The PCA (`PCA` function) on the Baccalaureate grades alone can be launched by choosing the default options. By default, all the available (quantitative) variables are active. This is why, in this command, the variables are restricted to columns 2 to 6.

```
> res=PCA(Bac[,c(2:6)])
```

The file `res` contains all of the results tables. By default, the basic graphs are displayed: representation of the individuals and variables on the first plane. We obtain other graphs using the `plot.PCA` function; for example, to obtain the representation of the variables on plane (3,4):

```
> plot.PCA(res,axes=c(3,4),choix="var")
```

To introduce the year grades (quantitative variables in columns 6 to 10) and the `high school` (qualitative variable in the first column) as supplementary:

```
> res=PCA(Bac,quanti.sup=c(7:11),quali.sup=1)
```

All of the qualitative variables must be explicitly declared as supplementary. The presence of an undeclared qualitative variable will lead to an error message. The individuals and the barycentres associated with the categories evolve within the same space and, by default, are represented simultaneously.

By default, the active and supplementary quantitative variables are represented simultaneously (see Figure 1.8, right). In order to obtain a graph featuring the barycentres of the categories alone (see Figure 1.8, left), we simply hide (`invisible`) the individuals.

```
> plot.PCA(res,axes=c(1,2),choix="ind",invisible="ind")
```

To group together the results tables in a single .csv file:

```
> write.infile(res,file ="Output_ACP_Bac.csv")
```

To obtain Table 1.3, we calculate the centred and reduced data table using the `scale` function (result in `Bac_CR`).

```
> Bac_CR=scale(Bac3[,2:11])*sqrt(909/908)
```

Remark

In the `scale` function, the variance used is the estimation of the population variance (denoted $\widehat{\sigma^2}$). Thus

$$\widehat{\sigma^2} = \frac{1}{I-1}\sum_i (x_i - \bar{x})^2 = \frac{I-1}{I}\text{Var}\,[x]\,.$$

To use the sample variance (denoted Var[x]), we therefore have to multiply the result of the `scale` function by $\sqrt{I/(I-1)}$.

This makes it possible to edit the centred-reduced data of individuals 46 and 238, for example:

```
> Bac_CR[c(46,238),]
```

Table 1.6 contains the transpose data (`t` function) rounded to two decimal places (`round` function) and limited to a set of individuals, the rank of which we put in a vector (called `list_ind_etiq`).

```
> list_ind_etiq=c(46,238,264,253,419,557,566,638,807,850,863)
> round(Bac_CR_t[,list_ind_etiq],2)
```

To obtain Table 1.4, we use the `catdes` function (category description) accessible in the **FactoMineR** menu, or using the command:

```
> results=catdes(Bac,num.var=1)
```

This function links a qualitative variable to each of the other variables in the file. Table 1.2 (right) is in:

```
> results$quanti.var
```

To obtain Table 1.2 (left), we apply the `catdes` function to the principal components. In order to do this, we concatenate (`cbind` function) the file of the individuals from the PCA (`res$ind$coord`) to the raw data:

```
> Tout=cbind(Bac,res$ind$coord)
```

We check the operation by listing the columns of the `data.frame` `Tout`.

```
> colnames(Tout)
[1]"HighSchool" "MAT" "PHY" "SN" "HG" "PHI" "mat3t" "phy3t" "sn3t"
[10] "hg3T"  "phi3T" "Dim.1" "Dim.2" "Dim.3" "Dim.4" "Dim.5"
```

Then we apply `catdes`, discarding the initial variables (from 2 to 11).

```
> results=catdes(Tout[,c(1,12:16)],num.var=1)
```

Table 1.4 (left) is in

```
> results$quanti.var
```

Script for Analysing the Orange Juice Data

Below is the script relating to Section 1.10. It contains some elements of R which are useful in factor analysis.

```
# Import and verification
> Orange=read.table("Orange5.csv",header=TRUE,sep=";",dec=",",
+ row.names=1)
> summary(Orange)

# PCA with the only active variables (Baccalaureate grades)
> res<-PCA(Orange[,1:18],quanti.sup=c(11:18),quali.sup=c(1,2))

# Figures 1.9 and 1.10
# Representation of individuals and then variables
# The cex argument modifies the font size of the labels
# The command x11() opens a new window

> plot.PCA(res,choix="ind",col.quali="black",cex=1.3)
> x11()
> plot.PCA(res,choix="var",cex=1.3,invisible="quanti.sup")
> x11()
> plot.PCA(res,choix="var",cex=1.3,invisible="var",
+ col.quanti.sup="black")
```

TABLE 1.6
Baccalaureate Data: Raw and Centred-Reduced Data for Several Students[a]

Raw Data											
Students	46	238	264	253	419	557	566	638	807	850	863
High School	14	7	9	9	8	17	15	8	6	14	21
Maths	10	3	19	17	15	11	10	18	20	18	5
Physics	7	9	17	15	18	1	6	15	12	16	1
Nat. Sci.	11	5	14	12	15	3	16	12	14	12	7
Hist.–Geo.	14	7	16	15	16	7	7	8	13	6	3
Philo	16	5	12	5	14	6	4	18	17	3	2
Maths year	8.25	7.5	16.92	13.93	13.53	7.83	8.5	15.2	15.47	14.33	6.6
Phys. year	7.17	9.77	14.6	13.96	14.6	5.83	8.83	15.43	15	11.6	5.67
Nat. Sci. year	9.8	10.3	16.17	15.33	12.33	7.67	11	13.25	14.27	12.9	7.67
Hist.–Geo. year	13	9.73	12.82	14.58	11.67	8.91	9	12.57	12.22	9.5	8.83
Philo year	9.83	7	10.4	8.1	11	6.83	6.67	11.08	12.07	7	10.33
Centred-Reduced Data											
Students	46	238	264	253	419	557	566	638	807	850	863
Maths	−1.01	−3.2	1.82	1.19	0.56	−0.69	−1.01	1.5	2.13	1.5	−2.57
Physics	−1.11	−0.55	1.66	1.11	1.94	−2.77	−1.38	1.11	0.28	1.38	−2.77
Nat. Sci.	0.01	−2.07	1.05	0.36	1.4	−2.77	1.74	0.36	1.05	0.36	−1.38
Hist.–Geo.	1.43	−1.21	2.19	1.81	2.19	−1.21	−1.21	−0.84	1.05	−1.59	−2.73
Philo	2.48	−0.86	1.26	−0.86	1.87	−0.56	−1.17	3.09	2.78	−1.47	−1.78
Maths year	−1.03	−1.31	2.27	1.13	0.98	−1.18	−0.93	1.62	1.72	1.29	−1.65
Phys. year	−1.48	−0.52	1.27	1.03	1.27	−1.97	−0.86	1.58	1.42	0.16	−2.03
Nat. Sci. year	−0.56	−0.33	2.38	1.99	0.61	−1.55	−0.01	1.03	1.5	0.87	−1.55
Hist.–Geo year	1.03	−0.66	0.93	1.85	0.34	−1.09	−1.04	0.8	0.62	−0.78	−1.13
Philo year	0.4	−1.11	0.7	−0.52	1.02	−1.2	−1.29	1.06	1.58	−1.11	0.66

[a] Identified in Figure 1.7.

2

Multiple Correspondence Analysis

Multiple correspondence analysis (MCA) is the factorial method adapted to tables in which a set of individuals is described by several qualitative variables. It can be presented in many different ways. In France, following the work of L. Lebart, the most common is to focus on the similarities with correspondence analysis, a method designed to study the relationship between two qualitative variables. In the perspective of simultaneously processing quantitative and qualitative variables for the same individuals, which is one of the strengths of multiple factor analysis (MFA), it is important to focus on the similarities between principal component analysis (PCA) and MCA. This is presented in this section. In the presentation chosen for this chapter, which follows that of PCA as closely as possible, we nonetheless do not come to the conclusion that MCA is simply a specific case of PCA. They are indeed two entirely distinct methods belonging to the same family: factorial analyses.

2.1 Data

The data are made up of a set of I individuals each described by a set of J qualitative variables. The structure of the raw data table is identical to that of the table used for PCA:

- Each row i corresponds to a (statistical) individual.
- Each column j corresponds to a (qualitative) variable.
- At the intersection of row i and column j, we find the value of variable j for individual i. Compared to PCA, the difference here is that this value is not quantitative: it is the category possessed by i for variable j (see Figure 2.1).

The most common example of this type of data, which will often serve as a reference, is that of a survey. The statistical individuals are the people questioned and the variables are the questions themselves. Thus, for the question, "What kind of job do you do?" participants respond by choosing from a set of predefined responses, for example: {manual labourer, employee, senior management, etc}.

In practice, in the questionnaires, the response categories for a given question are often ranked. Thus, in an opinion poll, a classic approach is to offer a set of statements (in surveys, it is called a *battery of items*): for each statement, the participant must express his or her agreement or disagreement using a scale identified by a set of ordinate categories. For example:

Express your agreement or disagreement with the following statement. All nuclear power plants should be closed.

1. Totally disagree
2. Disagree
3. Agree
4. Totally agree

We could consider variables like this to be quantitative: that is to say, a degree of agreement ranging from 1 (totally disagree) to 4 (totally agree). In an example, we show the advantage of considering them as qualitative.

2.2 Complete Disjunctive Table

The table described in the previous section corresponds to a table of data as it should be entered into software. Another way of representing the data is a table crossing the individuals in rows with the categories in columns featuring, at the intersection of row i and column k (belonging to variable j), value y_{ik} which is:

- 1 if individual i possesses category k (of variable j)
- 0 in all other cases

This table is not explicitly calculated by the user, but the MCA is based on it. It is known as a *complete disjunctive* table (and denoted CDT; see Figure 2.1): if we consider the y_{ik} relative to any individual and any variable, these values always contain one 1 (complete) and only one 1 (disjunctive). Furthermore, we denote:

- J the number of variables
- K_j the number of categories of variable j
- K the total number of categories: $K = \sum_j K_j$
- p_k the proportion of individuals possessing category k: $p_k = 1/I \sum_i y_{ik}$

Until now, we have implicitly considered that all the individuals are attributed the same weight ($1/I$, concretely, to obtain a total sum of weights equal to 1). This is indeed the case in the great majority of applications but

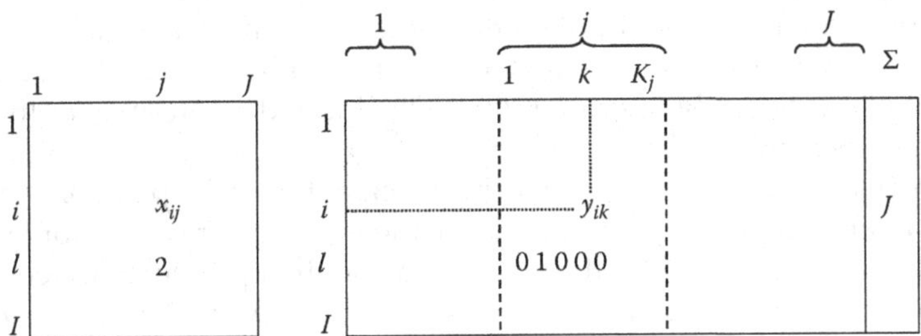

FIGURE 2.1
Condensed coding table (left) and complete disjunctive table (CDT, right). x_{ij}: Category of j possessed by i. $y_{ik} = 1$ if i possesses the category k (of j) and 0 in all other cases. Row l: Numerical example with $K_j = 5$ and $x_{ij} = 2$.

there is nothing to prevent us from attributing each individual a specific weight p_i. In this case, quantity p_k must account for them: $p_k = \sum_i p_i y_{ik}$

Quantity p_k can thus be interpreted as the weight of the set of individuals with category k. This justifies our choice to use the same letter p for the weight of an individual (p_i) as for a set of individuals (p_k). To simplify our presentation, we mostly examine cases in which the individuals have the same weights (thus considering p_k as a proportion).

The columns of this table are known as indicators (or indicator functions). They have a remarkable property which is highly important in MCA: the sum of the indicators for a given variable is a constant function (equal to 1). The result is that the column margin (a column for which the term i is the sum of terms for row i) is a constant function equal to J, the number of variables. The row margin, on the other hand, contains the category frequencies.

2.3 Questioning

Our example is a classic case: data from an opinion poll. The first step is to examine the frequencies in the different categories. First and foremost, the investigator will be interested in the answers to specific questions such as how many people like a given politician or how many students found a given course (statistics) useful, and so on. The second stage is to examine the relationships between pairs of variables. We might ask questions such as are favourite politicians linked to given professions or are attitudes to nuclear power related to attitudes to justice. The contingency table, constructed by crossing the responses to two questions, is the most common tool for analysing surveys and opinion polls.

These first two steps can be used to study the variability of respondents from both a one-dimensional and two-dimensional point of view. They are analogous with histograms and scatter plots for quantitative variables. Then, it is necessary to go further: as with PCA for quantitative variables, the main aim of MCA is to study the variability of the individuals from a multidimensional perspective.

With this in mind, we consider each participant from the point of view of all of his or her responses, which we can call their *response profile*. As for PCA, the diversity of these profiles will be studied using their principal dimensions of variability. At this general level, the objectives of PCA and MCA are identical.

Similarly, as in PCA, we are also interested in the relationships between the variables. However, the variables are qualitative: not all of a relationship can be summarised by an indicator such as the correlation coefficient, even in a first approach (the relationship indicators between two qualitative variables only measures the intensity or the significance of the relationship, but tells us nothing about the nature of the link). We must visualise the associations between categories: for example, the relationship between the variables *eye colour* and *hair colour* is characterised by the privileged association between the categories *blue eyes* and *blonde hair*; among other things, people with blue eyes more often have blonde hair than other people; blondes more often have blue eyes than other people. The most important elements of the results of an MCA lie in the graphical representation in which each category is represented by a point, and the proximity between two points expresses a privileged association between the two corresponding categories.

Also as in PCA, this overall visualisation of the relationships will be established from the synthetic quantitative variables used to construct the factorial planes. As in PCA, aside from their usefulness in representing categories, the search for synthetic variables can be an objective in itself. This perspective is less obvious in MCA as these synthetic variables are not of the same type as the variables that they synthesise. In concrete terms, a set of qualitative variables is summarised using quantitative variables by associating a coefficient to each category and, for each individual, calculating the sum of the coefficients for each of the categories it possesses. The difficulty is in choosing the coefficients. In factorial analysis, the main idea is to choose synthetic variables with the greatest possible variance.

2.4 Clouds of Individuals and Variables

Let us consider the complete disjunctive table. By following the approach used in PCA, we construct the cloud of rows (individuals) and of the columns (categories) for this table. The analogy with PCA is clearly seen when considering each column of the CDT as an indicator variable.

As in PCA, the data table must be transformed prior to the analysis. Indeed, in the raw CDT, the possession of a category k by an individual i induces the value $y_{ik} = 1$. However, this category k does not characterise individual i in the same way, depending on its frequency in the sample population: specifically, from the point of view of the analyst, possession of a rare category characterises an individual more than a frequent category. This is the origin of the idea behind relativising y_{ik} by the frequency of category k, which can be performed by transforming y_{ik} as follows:

$$x_{ik} \leftarrow y_{ik}/p_k.$$

With this transformation, possession of category k induces, for individual i, a value of x_{ik} equal to, for example:

- 1 if k is possessed by all individuals
- 2 if k is possessed by half the population
- 4 if k is possessed by a quarter of the population

Furthermore, the mean for each column k is worth 1. As the columns must be centred, the final transformation is expressed:

$$x_{ik} \leftarrow y_{ik}/p_k - 1.$$

In this chapter we present MCA from the unstandardised PCA of x_{ik}.

2.4.1 Cloud of Individuals

We consider all the data for individual i: $\{x_{ik}\ ;\ k = 1, K\}$. This individual can be represented by a point in space $\mathbb{R}^K$ (known as the individuals' space). We thus construct the cloud of individuals denoted N_I. By default, this cloud is centred (the origin is at the centre of gravity). Each individual i is attributed the weight p_i so $\sum_i p_i = 1$; this weight is generally constant, $p_i = 1/I$.

In defining distance in $\mathbb{R}^K$, we must specify which weight is attributed to each dimension, that is to say, to each category. In MCA, this weight is proportional to the frequency of the category, thus taking into account the constraint according to which the sum of all weights is 1: the weight of category $k = m_k = p_k/J$.

This counterbalancing is 'natural' if we consider a category as the set of individuals which possesses it, a valuable point of view for analysing the results, as we show later. The direct justification for counterbalancing in this way is clearer in the categories' space (the categories' weights in this space induce the metric within $\mathbb{R}^K$). Indirect justifications lie in the positive properties of the MCA.

Distance Between an Individual i and the Centre of Gravity of N_I

The centre of gravity of N_I is denoted G_I.

$$d^2(i, G_I) = \sum_k \frac{p_k}{J}\left(\frac{y_{ik}}{p_k} - 1\right)^2 = \frac{1}{J}\sum_k \frac{y_{ik}}{p_k} - 1.$$

This distance takes into account the not null coefficients y_{ik} for individual i, induced by the categories it possesses. The mean of these coefficients (more precisely its deviation from the value 1) indicates the extent to which the individual i presents rare categories. The more rare categories it presents, the farther it is from the origin (and thus the more unusual it is; a satisfactory interpretation of a distance from the origin).

Total Inertia of N_I (With Respect to G_I)

$$\text{Total Inertia } (N_I/G_I) = \sum_i p_i d^2(i, G_I) = \frac{K}{J} - 1.$$

The total inertia of the cloud depends not on the content of the data table but on one aspect of its format: the mean number of categories per variable. This result is the same as that of standardised PCA, in which the total inertia is equal to the number of variables (and therefore does not depend on the content of the table).

Distance Between Two Individuals i and l

$$d^2(i,l) = \sum_k \frac{p_k}{J}\left(\frac{y_{ik}}{p_k} - \frac{y_{lk}}{p_k}\right)^2 = \sum_j \frac{I}{J} \sum_{k \in K_j} p_k \left(\frac{y_{ik}}{p_k} - \frac{y_{lk}}{p_k}\right)^2.$$

The last term identifies the contribution of variable j to the distance between i and l. If individuals i and l present the same category for variable j, this contribution is worth 0. Otherwise, by denoting k (and, respectively, h), the category possessed by i (and, respectively, l), it has a value of

$$\frac{1}{J}\left(\frac{1}{p_k} + \frac{1}{p_h}\right).$$

Thus, two individuals are all the more distant when they possess different categories for a large number of variables, and particularly when these categories (thus possessed by only one of the two) are rare. This distance is more than satisfactory.

Remark

The distance between two individuals i and l can also be expressed:

$$d^2(i,l) = \sum_k \frac{J}{p_k}\left(\frac{y_{ik}}{J} - \frac{y_{lk}}{J}\right)^2 = \frac{1}{J}\sum_k \frac{1}{p_k}(y_{ik} - y_{lk})^2.$$

We reach this relationship when, as is usually the case, we present MCA from correspondence analysis: the data are transformed into profiles and the category k is attributed to the inverse of its weight. The PCA of x_{ik} does indeed yield the same results as the MCA.

2.4.2 Cloud of Categories

We keep in mind that a category, as a column of a CDT, can be considered as an indicator variable. After the transformation and centring introduced in the previous section, column k of the analysed table X is no longer exactly an indicator but differs only slightly as it is a function constant on the classes of the partition of the individuals associated with k (partition made up of two classes, that of individuals possessing k and that which groups the others together).

The categories, as columns of X, can thus be plunged into the space of functions on I (often known as the *variables' space* and denoted $\mathbb{R}^I$) identical to that introduced in PCA: each dimension corresponds to an individual; the weights of the individuals define the (diagonal) metric; the categories make up cloud N_K; being centred, they belong to the subspace orthogonal to constant functions. Each category is attributed a weight proportional to its frequency, thus for category k, p_k/J. This weight must be the same as that of the category in the definition of the distance in the individuals' space. In the following analysis, a direct justification of this weight is to favour the categories which concern a great number of individuals. The properties induced by these weights provide indirect justifications.

The categories are not reduced. The variance of category k, equal to the squared distance to the origin O due to the centring, is worth:

$$\text{Var}[k] = d^2(k, O) = \sum_i p_i \left(\frac{y_{ik}}{p_k} - 1\right)^2 = \frac{1}{p_k} - 1.$$

A category possessed by all of the individuals lies at the origin. Otherwise, the less frequently it occurs, the farther it is from the origin. However, in the factorial analysis of N_K, the categories intervene through their inertia. The influence of frequency p_k of category k thus intervenes on two levels:

1. That of the weight, which increases with p_k
2. That of the distance to the origin, which decreases with p_k

The inertia (with respect to O) of category k is worth

$$\text{Inertia}\,(k/O) = \frac{p_k}{J} d^2(k, O) = \frac{1 - p_k}{J}.$$

Finally, the rarer they are, the greater influence categories have. This result is fairly natural: in differentiating between individuals, very frequent categories are not of particular interest; very rare categories draw attention to the individuals who possess them. However, within the context of an overall study aiming to highlight general phenomena, focussing on a succession of individual cases is rather awkward. This is the reason for grouping together rare categories with others (for example, for an item such as that mentioned in

Section 2.1, if it is rarely used, we group together the extreme category *totally agree* with the category *agree*).

It must be noted, in passing, that the total inertia of N_K, obtained by adding together the inertias for all categories, is equal to that of N_I, which is an element of the duality between the two clouds shown in the case of quantitative variables. Thus,

$$\text{Inertia}\,(N_K/O) = \sum_k \text{inertia}(k/O) = \sum_k \frac{1-p_k}{J} = \frac{K}{J} - 1.$$

The following property is extremely important: the centre of gravity of the categories of one given variable lies at the origin of the axes. Indeed, the ith coordinate of this centre of gravity for variable j, is worth:

$$\sum_{k \in K_j} \frac{p_k}{J}\left(\frac{y_{ik}}{p_k} - 1\right) = \frac{1}{J}\left[\sum_{k \in K_j} y_{ik} - \sum_{k \in K_j} p_k\right] = \frac{1}{J}(1-1) = 0.$$

Thus, the centre of gravity of the whole of cloud N_K itself lies at the origin of the axes. Thus, in MCA, both clouds N_I and N_K are centred.

It seems helpful to evaluate the relative positions of categories, or in other words, the shape of cloud N_K. Following the PCA analogy suggests calculating the correlation coefficients between indicators. This unusual case of the correlation coefficient (called the point-biserial correlation coefficient) is not easy to interpret directly. Thus, in MCA, we are more interested in the distance between categories. When applied to categories k and h, this distance is expressed (where p_{kh} is the proportion of individuals who possess k and h simultaneously):

$$d^2(k,h) = \sum_i p_i \left(\frac{y_{ik}}{p_k} - \frac{y_{ih}}{p_h}\right)^2 = \frac{p_k + p_h - 2p_{kh}}{p_k p_h}.$$

In the last expression, the numerator represents the proportion of individuals who possess one, and only one, of the categories k and h. The denominator relativises (standardises?) this proportion by those of categories k and h taken separately (a same number of individuals possessing one and only one of the two categories k and h 'distances' yet the farther these two categories the rarer they are). The distance between two categories is therefore interpreted intuitively.

2.4.3 Qualitative Variables

Up until now, the qualitative variables have only been considered through their categories. It is clear that the categories play a central role in the MCA approach: an individual is characterised by the categories it possesses; the relationship between two qualitative variables is analysed through the associations between their categories. It is no less important to note that the variables as such influence interpretation.

In $\mathbb{R}^I$, variable j is first represented by its K_j categories. Prior to centring, the categories of a given variable are orthogonal two by two (the corresponding scalar products are null); they thus generate a subspace of K_j dimensions. This subspace is that of linear combinations of the indicators of j, therefore functions (defined) on I which are constant within the classes of the partition (of I) defined by j.

All of these subspaces share the axis of constant functions (for which the unit vector, which is made up only of ones, has already been denoted **1**). After centring (which consists of projecting N_K on the subspace orthogonal to **1**), the subspace associated with each variable j is of the dimension $K_j - 1$: it contains the centred functions constant within the classes of the partition defined by j.

The total inertia of the K_j categories k of variable j is worth:

$$\text{Inertia of categories of } j/O = \sum_{k \in K_j} \frac{1 - p_k}{J} = \frac{K_j - 1}{J}.$$

This inertia is even greater when the variable possesses many categories. Firstly, this result can be awkward as the user does not always control the number of categories of the variables; for example, the variable *gender* has two categories (man and woman) whereas the variable *region* has 21 categories (in France). Naturally, the user does not wish to make regions 20 times more important than gender. In fact, inertia $(K_j - 1)/J$ for variable j must be viewed in terms of the dimension $(K_j - 1)$ of the subspace generated by (the categories of) variable j: the higher the inertia, the more it is distributed according to a high number of dimensions. More precisely, as shown later, the inertia of the categories of variable j is constant (and equal to $1/J$) in the projection in whichever direction of the subspace they generate.

Thus, in researching the first axis of inertia, no one variable is favoured. However, the *gender* variable can only be strongly related to one single axis (therefore opposing men and women) whereas the *region* variable can be related to many (20) dimensions (opposing, for example, north and south, east and west, Brittany and Normandy, etc.). This fully justifies the proportionality of the total inertia of one variable with the number of categories.

If we project category k (of variable j) on a centred unit vector v of $\mathbb{R}^I$ (of which the ith coordinate v_i is the value of function v for individual i), the length of this projection is worth (where $\bar{v}_k$ is the average of function v for individuals possessing category k):

$$\langle k, v \rangle = \sum_i \frac{1}{I} \left(\frac{y_{ik}}{p_k} - 1 \right) v_i = \bar{v}_k.$$

This result is widely used in MCA and is also true when the individuals have different weights. Until now, we have considered a category essentially as a (indicator) variable. At this point, we encounter the second point of view, previously mentioned regarding the attribution of weights to categories: a category can be considered as the set of individuals which possess it. What is remarkable is that this second point of view, which naturally appears in the individuals' space, here appears in the space of functions (defined) on I.

Now, we project all of the categories k of variable j (denoted K_j) onto v. The inertia of these projections is worth:

$$\text{Projected inertia of } K_j \text{ on } v = \frac{1}{J}\sum_k p_k \bar{v}_k^2.$$

$$= \text{between-class inertia of } j = \frac{1}{J}\eta^2(j, v).$$

Up to the $1/J$ coefficient, this inertia is equal to the between-class inertia of the v function when considering the partition of individuals defined by variable j. As the total inertia of v is worth 1, this between-class inertia can be seen as the percentage of inertia (of v) 'explained' by qualitative variable j. It is thus equal to the squared correlation ratio between qualitative variable j and, on the other hand, quantitative function v. This indicator, denoted $\eta^2(j, v)$, is the classical measurement of the relationship between a quantitative variable and a qualitative variable. It can be shown that this indicator is interpreted geometrically as the squared cosine of the angle between quantitative variable v and the subspace generated by qualitative variable j.

This is a notable result. It shows that, within the same space of functions on I, it is useful to immerse both quantitative and qualitative variables as the measurement of (the intensity of) the relationship between two variables of different types is expressed simply in this space. This is the keystone of simultaneously analysing these two types of variables (as we do in factorial analysis of mixed data, FAMD, and of course in MFA); it offers a geometric framework for synthesising a set of qualitative variables by one (or a small number of) quantitative variable(s).

2.5 Fitting Clouds N_I and N_K

As in PCA, the two clouds N_I and N_K are each projected onto a series of orthogonal axes of maximum inertia.

2.5.1 Cloud of Individuals

Applying this procedure to the cloud of individuals is justified in the same way as in PCA: as cloud N_I is centred, the inertia of its projection in a given direction u measures the variance of the projected points. If we consider the coordinate of the projection of i on u as the value of i for the (linear) combination of initial variables defined by u, this inertia is the variance of this combination.

From a technical viewpoint, the only difference when compared with PCA is that, when calculating a distance in $\mathbb{R}^K$, the dimensions are attributed a weight (p_k/J for dimension k). In other words, space $\mathbb{R}^K$ is endowed with a (diagonal) Euclidian metric which is not the usual metric. If we organise these weights on the diagonal of a square matrix M, of dimension K, the scalar product between the two vectors u and v of $\mathbb{R}^K$ is expressed:

$$\langle u, v\rangle_M = u'Mv = \sum_k \frac{p_k}{J} u_k v_k.$$

We thus deduce:

$$\|u\|_M^2 = \langle u, u\rangle_M = u'Mu = \sum_k \frac{p_k}{J} u_k^2.$$

With these notations, along with those for PCA (see Section 1.5.1), the factor on I of rank s (vector for which the ith coordinate is that of the projection of i on u_s, denoted F_s) is expressed:

$$F_s = XMu_s.$$

The projected inertia of N_I on u_s is expressed (the weights p_i of the individuals are organised on the diagonal of the diagonal matrix D of dimension I):

$$F_s'DF_s = u_s'MX'DXMu_s.$$

The only real difference when compared with PCA is the presence of the weight of each dimension of $\mathbb{R}^K$ through matrix M. We are looking for a sequence of orthogonal axes with maximum inertia. For the direction of rank s, the criterion is expressed:

$$F_s'DF_s \text{ maximum}$$

with norm and orthogonality constraints:

$$\|u_s\|_M^2 = u_s'Mu_s = 1 \quad \text{and} \quad \langle u_s, u_t\rangle_M = u_s'Mu_t = 0 \quad \text{for } t < s.$$

It can be shown that vector u_s verifies

$$X'DXMu_s = \lambda_s u_s$$

with

$$\lambda_s = F_s'DF_s = u_s MX'DXMu_s.$$

As in PCA, the results of the MCA are based on diagonalisation. The eigenvectors define the axes on which cloud N_I is projected; the eigenvalues correspond to the projected inertias. Similarly to PCA, the F_s vectors are sometimes referred to as *principal components*.

2.5.2 Cloud of Categories

Again here, as in PCA, we project cloud N_K onto a sequence of orthogonal axes of maximum inertia. However, the properties of cloud N_K vary depending on which of the two analyses is used. In MCA, the projected inertia of categories K_j of a given variable j on a unit vector v is proportional to the squared correlation ratio between j and v. Therefore, the projected inertia of N_K is worth

$$\text{Projected inertia of } N_K \text{ on } v = \frac{1}{J}\sum_j \eta^2(j, v).$$

Thus, the factors (on I) of the MCA are the functions on I which are the most closely linked to the initial variables (in terms of the average of the squared correlation ratios). This property is highly important for two reasons:

1. It validates the process as a whole (application of the factorial methodology to the recoded CDT).
2. It draws yet another parallel between PCA and MCA (in PCA each factor on I, that is to say each principal component, maximises the sum of the squared correlation coefficients between itself and the initial variables).

Expressing eigenvalues as averages of correlation ratios makes their interpretation useful independently of the percentage of inertia. The maximum value 1 corresponds to a dimension shared by all of the subspaces created by the variables. In concrete terms, this dimension is associated with a synthetic partition (of individuals) in the sense that whatever the variable, if two individuals possess the same category, they belong to the same class of the synthetic partition. This is only possible if the qualitative variables are closely linked with one another: the categories of each variable can be merged so that the new J variables thus defined might be identical. The eigenvalue λ_s therefore measures the intensity of the relationships (between the variables) expressed by the axis of rank s.

The matrix formalisation of this approach in $\mathbb{R}^I$ uses that of the PCA (in $\mathbb{R}^I$) with, in addition, the weights of the columns taken into account (organised on the diagonal of matrix M). $G_s = X'Dv_s$ groups together the coordinates of the K categories along the axis of rank s (of unit vector v_s). We are looking for v_s which maximises the variance of these coordinates (in MCA, cloud N_K is centred thus making it possible to interpret inertia with respect to the origin as a variance; this is not the case in PCA) thus

$$G_s' M G_s = v_s' D X M X' D v_s$$

with norm and orthogonality constraints:

$$\|v_s\|_D^2 = v_s' D v_s = 1 \quad \text{and} \quad \langle v_s, v_t \rangle_D = v_s' D v_t = 0 \quad \text{for } t < s.$$

It is shown that v_s verifies

$$XMX'Dv_s = \lambda_s v_s$$

with $\lambda_s = G_s'MG_s = v_s'DXMX'Dv_s$.

Thus v_s is the unit eigenvector associated with the eigenvalue λ_s of $XMX'D$; the eigenvalues correspond to the projected inertias and are ranked in descending order.

2.5.3 Relationships Between the Two Analyses

As in PCA, the solution in $\mathbb{R}^K$ is expressed:

$$X'DXMu_s = \lambda_s u_s$$

that leads to

$$XMX'DXMu_s = \lambda_s XMu_s.$$

Thus, as $F_s = XMu_s$:

$$XMX'DF_s = \lambda_s F_s.$$

F_s, the sth factor on I, is, like v_s, an eigenvector of $XMX'D$ associated with the eigenvalue λ_s. These two vectors are collinear. The squared norm of F_s is λ_s and v_s is standardised. Thus

$$v_s = \frac{1}{\sqrt{\lambda_s}} F_s.$$

The linear combination of categories, as a vector of $\mathbb{R}^K$ on which N_I is projected (that is, u_s), yields a function on I (F_s) collinear to vector v_s of $\mathbb{R}^I$ on which N_K is projected.

This result clearly expresses the duality between the two analyses: the function on I the most closely related (in terms of the average of the squared correlation ratios) to all of the variables corresponds to the dimension of maximum variability of the cloud of individuals. This is the same as in PCA (if we replace *correlation ratio* with *correlation coefficient*). This is the basis for simultaneously interpreting the representations of N_I and N_K. In MCA, this duality is expressed in a particularly remarkable way. In calculating G_s, we express the link between v_s and F_s, thus

$$G_s = X'Dv_s = \frac{1}{\sqrt{\lambda_s}} X'DF_s.$$

For the kth coordinate, we obtain (as F_s is centred):

$$G_s(k) = \frac{1}{\sqrt{\lambda_s}} \frac{1}{p_k} \sum_i p_i y_{ik} F_s(i).$$

Therefore, up to a coefficient, the coordinate of category k on the axis of rank s (in $\mathbb{R}^I$) is equal to the mean of the coordinates (in $\mathbb{R}^K$) of the individuals possessing category k.

This property is vital to interpretation: the user is more likely to consider a category as a centre of gravity of the individuals than as a (projected) indicator. Concretely, the proximity on a factorial plane between categories k and h is interpreted as a profile similarity between the individuals possessing category h and those possessing category k. This similarity is automatic when these two categories are possessed by the same individuals (thus ensuring proximity between the corresponding indicators) but can be observed differently (the case of the two neighbouring categories of a given variable, for example, the age ranges 60–65 years and 65–70 years; no individual possesses both categories at once, but we might think that those which possess one have the same response profile as those possessing the other).

As in PCA, we use this reasoning by switching the roles of the rows and the columns.

In $\mathbb{R}^I$, vector v_s verifies

$$XMX'Dv_s = \lambda_s v_s.$$

Hence

$$X'DXMX'Dv_s = \lambda_s X'Dv_s.$$

But $G_s = X'Dv_s$. Therefore

$$X'DXMG_s = \lambda_s G_s.$$

G_s is an eigenvector of $X'DXM$ associated with the eigenvalue λ_s and is therefore collinear to u_s,

$$G_s = \sqrt{\lambda_s} u_s.$$

This property is used to express F_s according to G_s. Thus

$$F_s = XMu_s = \frac{1}{\sqrt{\lambda_s}} XMG_s.$$

Hence, for the coordinate of row i along u_s (G_s is centred):

$$F_s(i) = \frac{1}{\sqrt{\lambda_s}} \frac{1}{J} \sum_k y_{ik} G_s(k).$$

Thus, up to a coefficient, the coordinate of individual k on the axis of rank s (in $\mathbb{R}^K$) is equal to the mean of the coordinates (in $\mathbb{R}^I$) of the categories it possesses. This relationship and the previous one (expressing $G_s(k)$ according to $F_s(i)$) are known as *transition relations* as they express the coordinates of the points in one space ($\mathbb{R}^I$ or $\mathbb{R}^K$) according to the coordinates of the points in the other space.

2.6 Representing Individuals, Categories and Variables

Representing Individuals and Categories

In MCA, transition relations are particularly simple and suggest superimposing the representations of individuals and categories. They can therefore be expressed as follows. Along each axis, up to a coefficient:

- An individual is at the barycentre of the categories it possesses.
- A category is at the barycentre of the individuals that possess it.

This is why the term *barycentric properties* is used for transition relations (we also say *quasi-barycentric* to refer to the fact that the categories are barycentres only up to a coefficient).

In the practice of MCA, for example, in opinion poll data processing, the individuals are often numerous and anonymous (the only thing we know about them are their answers). Thus, the cloud of individuals is only consulted for its general appearance, with attention focused on the representation of the categories. With this in mind, some users prefer to represent categories by their exact barycentres.

Representing the Variables

Particularly when there are a high number of variables, considering the variables is useful prior to considering the categories. Some questions users might want to know the answers to are which variables are most closely related to a given axis, and which variables contributed most to a given axis.

Thanks to the correlation ratio (between variable j and the factor on I of rank s) it is possible to answer these two questions as its square (already denoted $\eta^2(j, F_s)$) measures both:

- The intensity of the relationship between variable j and the factor of rank s
- The contribution of variable j to the inertia of dimension s (see expression of this inertia according to $\eta^2(j, F_s)$)

These coefficients can be represented by a graph in which the coordinate of variable j along direction s is $\eta^2(j, F_s)$. Such a graph, an example of which is given in Figure 2.3, highlights those variables most closely linked to one of the two axes or to both of them. As the coordinates are between 0 and 1, the points are situated within a square (with sides with a length of 1), which is why this graph is called the *relationship square*. It must be noted that we can also represent quantitative variables on this graph by using the squared correlation coefficients as coordinates (between these variables and the factors F_s). Other properties of this relationship square are described in the context of MFA (Chapter 7).

Number of Axes
In $\mathbb{R}^K$, it is possible to find K orthogonal axes. But, in MCA, the K_j categories of variable j are linked by a relationship (their sum counterbalanced by the weights of the categories is null). From this point of view, the maximum number of axes is $K - J$. In addition, cloud N_I is made up of I points; at most $I - 1$ axes are necessary in order to represent them perfectly. Finally, in MCA, the maximum number of axes with not-null inertia is $\min\{I - 1,\ K - J\}$.

In $\mathbb{R}^I$, the categories, being centred, are situated within a subspace of dimension $I - 1$. Furthermore, after centring, each variable j generates a subspace of dimension $K_j - 1$. All of the variables therefore generate a subspace with a maximum dimension of $K - J$. Finally, again we find that in MCA, the maximum number of axes with nonzero inertia is $\min\{I - 1,\ K - J\}$.

2.7 Interpretation Aids

The interpretation aids presented for PCA (Section 1.6) can be directly applied to MCA.

Percentage of Inertia Associated with an Axis
Inertia itself (associated with an axis), as a mean of squared correlation ratios, has already been discussed. The percentage of inertia, as in all factorial analysis, measures the representation quality of the whole cloud (N_I or N_K) by an axis or plane.

In practice, compared to PCA, MCA generally leads to:

- Smaller percentages of inertia
- A smoother decrease of these percentages

When faced with such a result, users accustomed to working with PCA are often disappointed, in particular by the low percentages of inertia. This point requires some clarification. Everything stems from the fact that each qualitative variable (with K_j categories) corresponds to a subspace (of K_j dimensions). We can get a feeling for these low percentages by considering the limit case in which all the variables are identical to one another. In such a case:

- PCA yields a first axis associated with a percentage of inertia of 100% (and thus 0 for all following axes).
- MCA yields $K_j - 1$ axes across which the 100% of inertia is evenly distributed.

Thus, for example, in the presence of variables with five categories, the percentage of inertia associated with the first axis cannot exceed 25%. More generally, we can calculate the ratio between the upper limit of projected inertia (1) and the total inertia ($K_j - 1 = 4$ in this example).

Contributions

The same comments can be made about the contribution of an individual in MCA as in PCA. The contributions of the variables have already been discussed. The contribution (denoted $CTR_s(k)$) of a category k (to the inertia of axis of rank s) can be considered in two different ways. From its definition in space $\mathbb{R}^I$,

$$CTR_s(k) = \frac{p_k}{J} G_s(k)^2,$$

in $\mathbb{R}^K$ we obtain (by denoting $\bar{F}_s(k)$ the coordinate of the barycentre of the individuals possessing the category k):

$$CTR_s(k) = \frac{1}{J}\frac{1}{\lambda_s} p_k \bar{F}_s(k)^2.$$

The contribution of a category k can therefore also be expressed in $\mathbb{R}^K$: it is proportional to the inertia of the barycentre of the individuals possessing category k. This point of view enhances the interpretation of the MCA criterion in $\mathbb{R}^K$. Until now, we have aimed to express the maximum variability of the individuals by a reduced number of dimensions. If, within $\mathbb{R}^K$, we consider the cloud of barycentres associated with the categories, it appears that the axes of the MCA also separate these barycentres 'at best'.

Supplementary Elements

In practice, as for PCA, MCA is almost always performed using supplementary elements. Introducing supplementary individuals or qualitative variables (and therefore categories) is not particularly problematic. Apart from the fact that they do not participate in constructing the axes, these elements are processed in the same way as the active elements.

Moreover, we can introduce supplementary quantitative variables in MCA. In such cases, we calculate the correlation coefficients between these variables and the factors on I. As in PCA, the variables are then represented in the correlation circle. They can also be represented in the relationship square (see Section 2.6).

2.8 Example: Five Educational Tools Evaluated by 25 Students

2.8.1 Data

For a first look at the method, it is best to apply it to a small dataset which will lead to clear interpretations. In order to do this, we constructed a small table

TABLE 2.1
Educational Tools. Raw data

N°Ind.	Text	Animation	Films	Class Book	Exercise Book
1	1	1	1	1	1
2	2	2	1	1	2
3	1	2	2	2	1
4	2	2	2	2	1
5	3	2	3	1	2
6	1	1	1	5	4
7	2	1	1	4	4
8	3	3	3	1	2
9	3	3	3	2	1
10	2	2	2	4	3
11	1	1	2	5	5
12	1	2	1	5	5
13	3	3	3	3	3
14	3	4	4	2	2
15	3	4	5	2	1
16	5	4	4	1	1
17	4	4	4	2	2
18	3	3	3	5	4
19	4	5	5	2	2
20	4	4	4	3	4
21	4	4	4	4	4
22	5	5	5	4	4
23	5	4	5	5	5
24	5	4	5	5	5
25	5	5	4	5	5

which is the result of a survey of 25 students about their opinion concerning the usefulness of educational tools available to them.

There are five tools: three of them are components of an online class, that is to say a text, animations and films describing software use. There are also two books, one a class textbook and the other a book of exercises. We asked students to evaluate the usefulness of each of these five tools on a five-point scale ranging from 1 = useless to 5 = very useful. The raw data table is made up of $I = 25$ rows and $J = 5$ columns (see Table 2.1). Here the variables are considered to be qualitative; the table is therefore analysed using MCA.

The categories of these variables are ordered. As previously mentioned, this type of variable (often known as an *ordinal variable*) can also be considered quantitative and can therefore be analysed by a PCA. At the end of this chapter, we will make some comparisons between these two points of view, using this dataset.

Finally, it must be noted that, in practice, the number of individuals in this study (25) is not sufficient for performing MCA. This is because studying

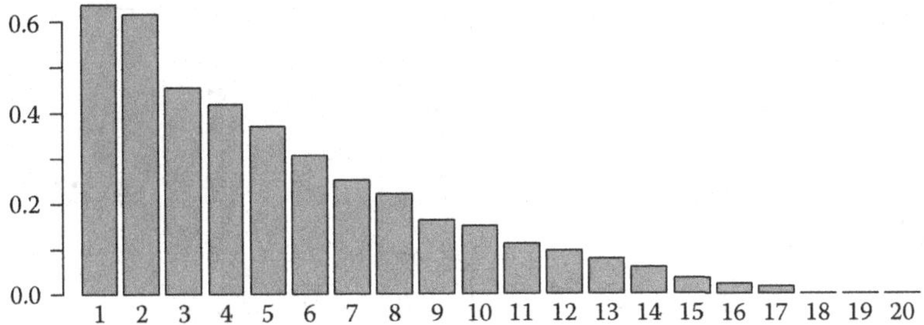

FIGURE 2.2
Ed. Tools. MCA. Eigenvalues.

relationships between qualitative variables requires more individuals than when studying relationships between quantitative variables. To get a good idea of this, let us consider the contingency table confronting two variables. In the example, no matter which pair of variables we are considering, the table has 25 cells. It is clear that 25 individuals are not sufficient to identify 'stable' irregularities in the distribution of these individuals over the 25 cells (it is these irregularities that show the notable associations between categories, which are at the heart of the relationship between qualitative variables). That being said, as our objective is to illustrate the method by connecting the results of an MCA to the raw data, this 'instability' is not problematic (especially as the data chosen to illustrate the MCA do not result from a real survey).

2.8.2 Analyses and Representations

The decrease of the eigenvalues (see Figure 2.2) clearly suggests retaining the first two axes for interpretation. The first two eigenvalues are similar: the first plane formed by the first two axes is stable rather than the axes themselves.

The percentages of inertia associated with these first two axes (15.92% and 15.42%) may seem low to PCA users. However, when we consider the number of categories for each variable, we know that this percentage cannot exceed 25%. As the first two eigenvalues themselves (0.64 and 0.62) are means of squared correlation ratios, they can be considered as high: each of the first two factorial dimensions is closely related to all of the variables.

If we decompose this inertia according to the variables, we obtain (up to the coefficient $J = 5$) the squared correlation ratios between the factors on I and the variables (see Figure 2.3). We can clearly see that the first two axes are dominated by the three components of the online classes and that the two books only intervene in forming the first axis.

Representing Individuals

The representation suggests a tripolar structure (see Figure 2.3, right); the first axis separates individuals {5, 8, 9, 13, 18} from the others. The second axis separates the remaining individuals into two distinct groups.

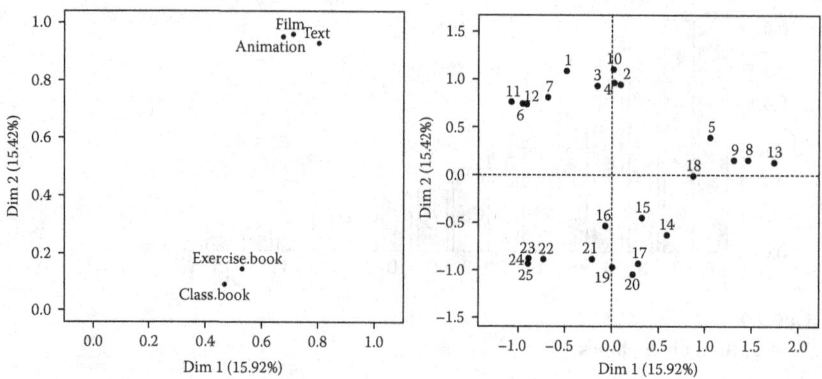

FIGURE 2.3
Ed. Tools. MCA, first plane. Relationship square (left). Representation of individuals (right).

Representing Categories

As the categories of a given variable are ordered, we highlight this information by connecting them (in order) on the graph (the broken line thus obtained is often called the *trajectory*).

We begin by commenting on axis 2, which is the simplest (see Figure 2.4). Indeed it highlights a 'linear' relationship between the three components of the online class: axis 2 organises the students from those who judged these three components as useless (example: n°1) to those who judged them to be very useful (example: n° 22). This dimension, which we could summarise as being the attitude to online classes, is not related to opinions about the books.

The first axis, however, opposes:

- For the online class, average categories with extreme categories
- For the books, categories *quite useful* and *very useful* with the others

As is (almost) always the case, it is concrete to reason in terms of the individuals (this is particularly true in this example where the distribution of individuals is discontinuous). The first axis therefore identifies individuals 5, 8, 9 and 13 who expressed average opinions about the online classes and negative or average judgements about the books. Opposite, we find the individuals who found the books very useful but whose opinions were divided concerning the online class, with some finding it useful (21 to 25) whereas the others did not (6, 7, 11 and 12).

Despite its small size, this example clearly illustrates the results of the MCA that the users need: a representation of the individuals emphasising the principal dimensions of their variability, a representation of categories emphasising their most remarkable associations and a tool for describing the relationships between qualitative variables.

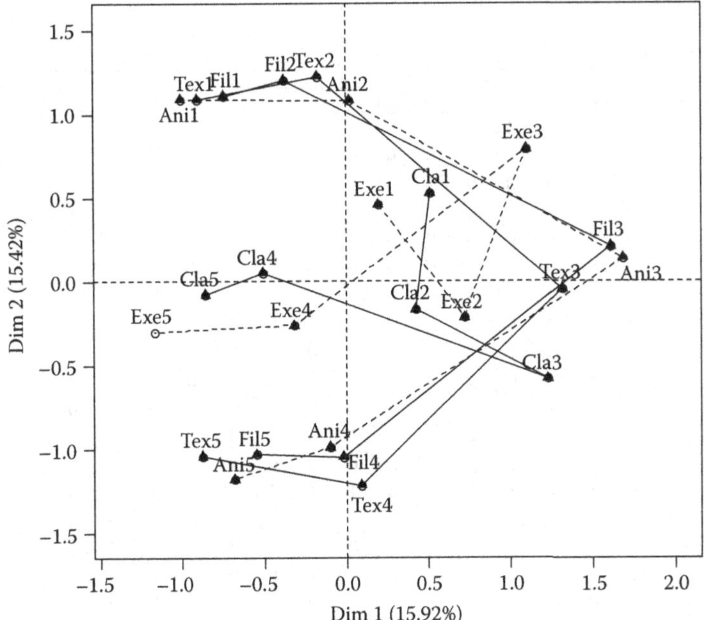

FIGURE 2.4
Ed. Tools. MCA. Representation of categories on the first plane.

2.8.3 MCA/PCA Comparison for Ordinal Variables

The categories of the variables in these data are ordered. As we have already said, this characteristic is sufficiently common to warrant further comment; in particular, such variables can also be considered as quantitative and thus processed using a PCA. This example is the opportunity to compare these two approaches.

PCA only considers linear relationships. In the space of functions on I ($\mathbb{R}^I$), a quantitative variable is represented by a vector; the relative positions of two variables is limited to their angle (the cosine of which geometrically represents the correlation coefficient). In MCA, qualitative variables are represented by the same number of points as of categories. The relative positions of two variables can be of various patterns. This corresponds well to the notion of relationships between qualitative variables which are much more complex, or from another point of view, much richer, than the linear relationships between quantitative variables.

In MCA, emphasising nonlinear relationships is characterised in concrete terms by:

- Irregular deviations between consecutive categories in a general monotone pattern; this is the case for the components of the online class of which categories 2 and 4 are much closer to the corresponding extreme categories (1 and 5) than to the central category 3; this

TABLE 2.2
Ed. Tools. Burt Table[a]

	Tex1	Tex2	Tex3	Tex4	Tex5	Ani1	Ani2	Ani3	Ani4	Ani5	Fil1	Fil2	Fil3	Fil4	Fil5	Lco1	Lco2	Lco3	Lco4	Lco5
Ani1	3	1	0	0	0	4	0	0	0	0										
Ani2	2	3	1	0	0	0	6	0	0	0										
Ani3	0	0	4	0	0	0	0	4	0	0										
Ani4	0	0	2	3	3	0	0	0	8	0										
Ani5	0	0	0	1	2	0	0	0	0	3										
Fil1	3	2	0	0	0	3	2	0	0	0	5	0	0	0	0					
Fil2	2	2	0	0	0	1	3	0	0	0	0	4	0	0	0					
Fil3	0	0	5	0	0	0	1	4	0	0	0	0	5	0	0					
Fil4	0	0	1	3	2	0	0	0	5	1	0	0	0	6	0					
Fil5	0	0	1	1	3	0	0	0	3	2	0	0	0	0	5					
Lco1	1	1	2	0	1	1	2	1	1	0	2	0	2	1	0	5	0	0	0	0
Lco2	1	1	3	2	0	0	2	1	3	1	0	2	1	2	2	0	7	0	0	0
Lco3	0	0	1	1	0	0	0	1	1	0	0	0	1	1	0	0	0	2	0	0
Lco4	0	2	0	1	1	1	1	0	1	1	1	1	0	1	1	0	0	0	4	0
Lco5	3	0	1	0	3	2	1	1	2	1	2	1	1	1	2	0	0	0	0	7
Lex1	2	1	2	0	1	1	2	1	2	0	1	2	1	1	1	2	4	0	0	0
Lex2	0	1	3	2	0	0	2	1	2	1	1	0	2	2	1	3	3	0	0	0
Lex3	0	1	1	0	0	0	1	1	0	0	0	1	1	0	0	0	0	1	1	0
Lex4	1	1	1	2	1	2	0	1	2	1	2	0	1	2	1	0	0	1	3	2
Lex5	2	0	0	0	3	1	1	0	2	1	1	1	0	1	2	0	0	0	0	5

[a] juxtaposition of tables crossing variables pairwise.

'isolation' of category 3 stems from the fact that, for the components of the online classes, these categories are almost systematically associated with one another (in comparison, category 1 is associated with 2 and 4 with 5).

- Convergences between extreme categories (1 and 5) of a variable which are associated in the same way with the categories of another question; this, for example, is the case of *exercise_book_5* (Exe5) which is associated as much with category 1 as for 5 for the text component of the online class.

We can easily find the origins of these representations by examining tables crossing variables pairwise (see Table 2.2).

The representations of individuals and variables on the first plane of the PCA performed on these same data (see Figure 2.5) show:

- A strong correlation between the components of the online class; the students who find one component useful (and, respectively, useless) generally find the other components useful (and, respectively, useless).
- A strong correlation between the two books; the students who find one book useful (and, respectively, useless) generally find the other book useful (and, respectively, useless).

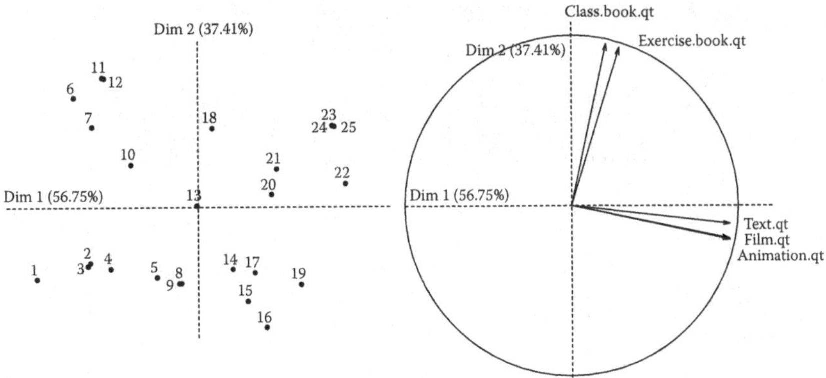

FIGURE 2.5
Ed. Tools. PCA. Representation of individuals (left) and variables (right).

– No correlation between the online class and the books.

This example clearly illustrates that which we can expect from the two approaches. As expected, the linear relationships between the variables appear more clearly in PCA, but the nonlinear aspects are absent.

It should be noted that the percentage of inertia associated with the first plane varies considerably from one analysis to the other (31.33% in MCA; 94.16% in PCA). PCA users may believe they have accurately described the variability of their data, but this will not be the case for those using MCA. However, the MCA plane is richer than that of PCA. This is because in PCA we are limited to linear relationships and in this case the first plane of this PCA gives a near-perfect image of these data. The context of MCA is wider-reaching. This can be illustrated with a simple example: it is better to have 31.33% of four million euros than 94.16% of only one million (1 and 4 are the dimension of the subspace generated by one variable in each of the two cases).

However, the plane of the PCA is valuable precisely because of its minimalist nature. We show in Section 8.4 how MFA can provide a method uniting the advantages of both approaches.

2.9 MCA in FactoMineR

To illustrate this application, we use the Educational tools data. The variables are introduced twice, both as qualitative and as quantitative. In order to work with a more general dataset, a qualitative variable (which is supplementary) is added (the discretised sum of the five grades called Class-Grade: A, B, C, D, E).

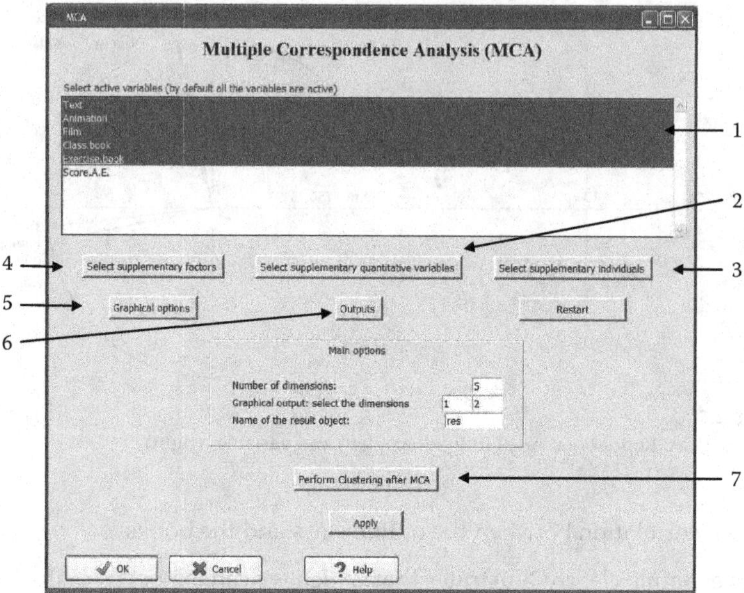

FIGURE 2.6
Main MCA window in the drop-down **FactoMineR** menu.

Drop-Down Menu in R Commander

The window which appears when we select MCA is reproduced in Figure 2.6. The comments are the same as those made for PCA.

- 1. (Qualitative) active variables are selected in the main window. They are not necessarily adjacent in the table. If nothing is selected, all of the (qualitative) variables are active and the quantitative variables are ignored.
- 2, 3 and 4. By default there are no supplementary elements. By using the drop-down menu, we ignore the variables in the file which are not selected as active or supplementary.
- 5. Open the window as seen in Figure 2.7.
- 6. Select the results tables to be listed and specify a .csv file name in which all of these tables will be brought together.
- 7. Used to conduct a clustering of individuals (Ward's method) from factorial coordinates, this clustering following a factorial analysis is available for all the factorial methods. It is illustrated for MFA in Chapter 4.

FactoMineR offers three types of graph for MCA (see Figure 2.7).

1. This is the classic graph for MCA which can contain both the individuals and the categories of the qualitative variables (known in R as *factors*), no matter whether they are active or supplementary.

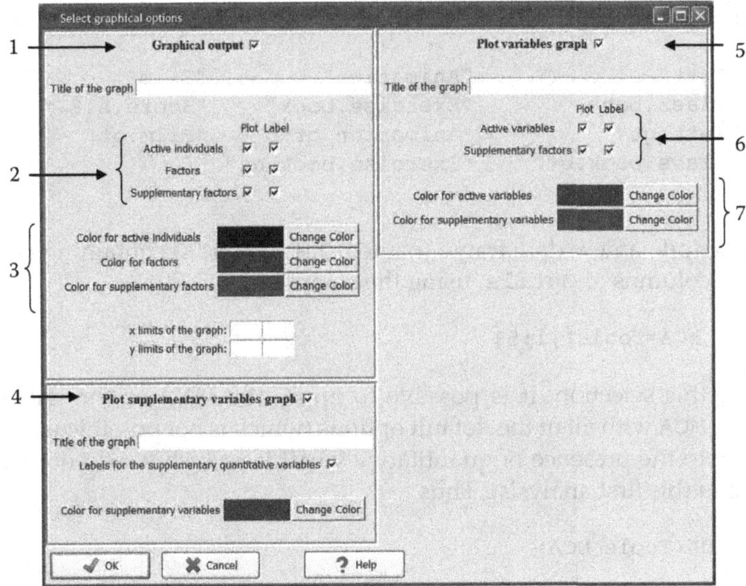

FIGURE 2.7
MCA Graphical options window.

2. These three types of elements can be selected and labelled (or not) independently. Often, it is not useful to label the individuals, and the general shape of the cloud they form is sufficient.
3. Attribution of colours to the types of elements to be represented.
4. The quantitative variables, always supplementary in MCA, are represented using their correlation coefficients with the factors. The result is a correlation circle which is read in exactly the same way as in PCA for the supplementary variables.
5. The relationship square (see Figure 2.3) is used to represent the qualitative variables themselves simultaneously (via their correlation ratio with the factors), rather than their categories, along with the quantitative variables (via their correlation coefficient).

6 and 7. The different types of variables can be individually selected, labelled and coloured.

Examples of Commands

The data are in the file `E_tools.csv`. For the sake of verification, the names of the variables are listed in the order of the file, by:

```
> Tools=read.table("E_tools.csv",sep=";",header=TRUE,
+ row. names=1)
```

```
> colnames(Tools)

 [1] "Text"               "Animation"         "Film"
 [4] "Class.book"         "Exercise.book"     "Score.A.E."
 [7] "Text.qt"            "Animation.qt"      "Film.qt"
[10] "Class.book.qt"      "Exercise.book.qt"
```

We compile a new data.frame (named `Outils_MCA`) containing only the five first columns of `Outils`, using the command

```
> Tools_MCA=Tools[,1:5]
```

Using this selection, it is possible to apply the MCA to the data.frame `Outils_MCA` with all of the default options (which is not possible with `Outils` due to the presence of quantitative variables which we do not want to include in this first analysis). Thus

```
> res=MCA(Tools_MCA)
```

This command is used to display basic graphs (individuals, categories and variables on the first plane) and generates all of the results tables (coordinates, contributions, and so on, for the individuals, categories and so forth) in the `res` list. This is an `MCA`-class list, an attribute recognised by the generic `plot` function. In practice, there is therefore no difference between creating the graphs using the `plot.MCA` function introduced below, and the generic `plot` function (as long as the **FactoMineR** package has been loaded, of course). All of these tables can be exported in a .csv file using the following command.

```
> write.infile(res,file ="Output_MCA_Tools.csv")
```

It is also possible to specify the columns to be selected in the `MCA` function.

```
> res=MCA(Tools[,1:5])
```

The graphs are created using a function specific to MCA: `plot.MCA`. This function constructs the type of graph required to suit the elements to be represented, elements defined by the `choix` argument, thus:

1. `Choix="ind"`. Usual graph: individuals + categories (active and supplementary)
2. `Choix="quanti.sup"`. Correlation circle for the quantitative variables (which must be supplementary)
3. `Choix="var"`. Relationship square

By default, all of the selected elements (individuals and variables) are labelled. Thus, the relationship square (see Figure 2.3 on the left) is obtained by

```
> plot(res,choix="var")
```

Elements can be removed from the selection using the argument `invisible`; the graph of individuals alone (see Figure 2.3 on the right) is obtained by

```
> plot.MCA(res,choix="ind",invisible="var")
```

The parameter `label` can be used to label only certain elements; to obtain the graph of labelled categories, when the individuals are not labelled:

```
> plot.MCA(res,choix="ind",label="var")
```

Now let us apply the MCA to the entire `Tools` table. The qualitative supplementary variables must be identified as such (otherwise they are considered active) as must the quantitative supplementary variables (otherwise they generate an error message).

```
> res=MCA(Tools,quali.sup=6,quanti.sup=c(7:11))
```

By default, this command displays all of the basic graphs: individuals (`ind`), categories (`var`), quantitative variables (`quanti.sup`; correlation circles) and all of the variables (`var`; relationship square).

The `plot.MCA` function has many options. For example, representing the individuals alone, coloured according to the (qualitative) variable n°6, using the argument `habillage` (`hab=6`), and so on:

```
> plot.MCA(res,choix="ind",invisible=c("var","quali.sup"),
+ hab=6)
```

The following command displays the relationship square (`choix = "var"`) for active variables alone (`invisible = c("quanti.sup", "quali.sup")`).

```
> plot.MCA(res,choix="var",invisible=c("quanti.sup","quali.
+ sup"))
```

It can often be handy to represent the eigenvalues as a barplot. Figure 2.2 was obtained using the command:

```
> barplot(res$eig[,1],names=1:20,cex.names=0.8)
```

The eigenvalues are in the first column of `res.mca$eig`. The `names` argument contains the names of the bars (here the rank of the axis) whereas `cex.names` changes the size of the letters for the bar names (here, this is reduced so as to display the name of each bar without creating an excessively large graph). If we only want to display the first ten eigenvalues:

```
> barplot(res.mca$eig[1:10,1],names=1:10)
```

Figure 2.4 is obtained by displaying an active window with the positions of the points via `plot.MCA`. In this window, we add the lines (using the `points` function).

In the `res` list produced by the MCA, the coordinates of the categories are in `res$var$coord`. These categories are ranked according to the variable and, within each variable, in alphabetical order. In this example, this means ranking them by degree of usefulness in ascending order (from `fil1`–useless films – to `fil5` – very useful films). In order to link the categories of a given variable in their natural order, we indicate their coordinates in the `points` function and use the argument `type = "o"` to display the points and to connect them. Thus, to connect the categories of the first variable (Text):

```
> plot(res,choix="ind",invisible="ind")
> points(res$var$coord[1:5,1],res$var$coord[1:5,2],type="o")
```

To connect those of the second variable (rows 6 to 10):

```
> points(res$var$coord[6:10,1],res$var$coord[6:10,2],type="o")
```

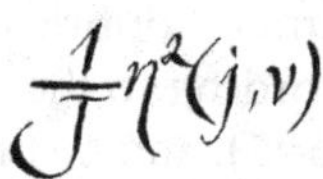

3

Factorial Analysis of Mixed Data

The need to introduce simultaneously quantitative and qualitative variables (known as mixed data) as active elements of one factorial analysis is common. The usual methodology is to transform the quantitative variables into qualitative variables, breaking down their variation interval into classes, and submitting the resulting homogeneous table to a multiple correspondence analysis (MCA). This methodology is relatively easy to implement and is used whenever there are enough individuals; generally more than 100, a limit below which MCA results are not very stable.

In two cases, there are advantages to conserving the quantitative variables:

1. When the number of qualitative variables is very low compared to quantitative variables: thus, we might think twice about recoding 20 quantitative variables with the sole aim of introducing one single qualitative variable
2. When there are only a small number of individuals

The method we present here stems from two different origins. In 1979, Brigitte Escofier suggested introducing quantitative variables in MCA (thanks to appropriate coding). In 1990, Gilbert Saporta suggested introducing qualitative variables in principal component analysis (PCA) thanks to a specific metric. In reality, these two different approaches yield the same results. The resulting factor analysis presents a sufficient number of positive properties and application potential to justify the status of a separate method: factorial analysis of mixed data (FAMD).

3.1 Data, Notations

We have I individuals. Each individual i is attributed a weight p_i such as $\sum_i p_i = 1$. To simplify matters, except when explicitly stated, we suppose that the individuals are of the same weight, thus $p_i = 1/I \ \forall i$. These individuals are described by:

- K_1 quantitative variables $\{k = 1, K_1\}$; these variables are standardised (centred and reduced); this is not merely for convenience but is necessary due to the presence of two types of variables.

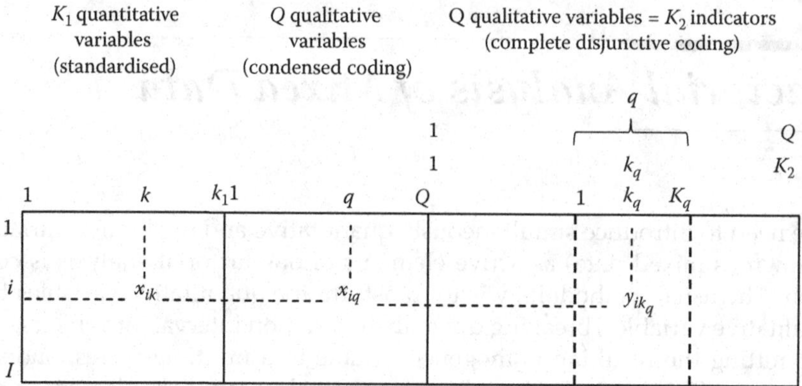

FIGURE 3.1
Data structure and principal notations. x_{ik}: Value of i for variable (centred-reduced) k; x_{iq}: Category of i for variable q; y_{ik_q}: 1 if i possesses k_q of variable q and otherwise 0.

- Q qualitative variables $\{q = 1, Q\}$; the qth variable presents K_q categories $\{k_q = 1, K_q\}$; the overall number of categories is $\sum_q K_q = K_2$; we denote p_{k_q} the proportion of individuals possessing category k_q.

Let $K = K_1 + K_2$ the total number of quantitative variables and indicator variables.

These notations can be brought together in the table in Figure 3.1 in which the qualitative variables appear both in their condensed form and in their complete disjunctive form.

3.2 Representing Variables

Let $\mathbb{R}^I$ be the space of functions on I. This space is endowed with the diagonal metric of the weights of the individuals, denoted D:

$$D(i, j) = \begin{cases} 0 & \text{if } j \neq i \\ p_i & \text{if } j = i \end{cases}$$

Generally the individuals have the same weights: $D = (1/I)\, I_d$ (where I_d is the identity matrix of appropriate dimensions).

As in standardised PCA, the quantitative variables are represented by vectors with a length of 1.

As in MCA, variable q is represented by cloud N_q of its centred indicators K_q. This cloud generates subspace E_q of dimension $K_q - 1$; E_q is the set of centred functions constant on the classes of the partition defined by q. For N_q to possess the same inertial properties as in an MCA, if we perform an unstandardised PCA on it, the indicator k_q must be divided by p_{k_q} and attributed a weight p_{k_q} (strictly speaking, obtaining the exact inertia of the MCA requires the weight p_{k_q}/J (see Section 2.4.1). Dividing by J 'averages' the inertias according to the number of variables, which is undesirable here as the qualitative variables are confronted with quantitative variables the inertias of which are not averaged).

Specifically, by proceeding in this way, we obtain a fundamental property of MCA: the projected inertia of N_q on a centred variable y is equal to the squared correlation ratio $\eta^2(q, y)$ between q and y.

When looking for direction v of $\mathbb{R}^I$ which maximises the projected inertia of cloud N_K (made up of the quantitative variables and the indicators), we maximise the criterion:

$$\sum_{k \in K_1} r^2(k, v) + \sum_{q \in Q} \eta^2(q, v).$$

This is the starting point of the method put forward by Gilbert Saporta in 1990. Geometrically, as variables k are standardised, the projection coordinate of variable k on v is worth $\cos(\theta_{kv}) = r(k, v)$, where θ_{kv} is the angle between vectors k and v. Similarly, as v is centred, $\eta^2(q, y) = \cos^2(\theta_{qv})$ where θ_{qv} is the angle between v and its projection on E_q. The criterion is thus expressed:

$$\sum_{k \in K_1} \cos^2 \theta_{kv} + \sum_{q \in Q} \cos^2 \theta_{qv}.$$

This is the starting point of the method put forward by Brigitte Escofier in 1979.

The influence of a variable must be explained according to the dimension of the subspace it generates. Thus, in space $\mathbb{R}^I$:

- A quantitative variable is represented by a vector associated with an inertia of 1.
- A qualitative variable with K_q categories is represented by K_q vectors generating a subspace E_q of dimension $K_q - 1$, all of which are associated with an inertia of $K_q - 1$.

As in MCA the total inertia of a qualitative variable increases the more categories there are. However, when projected on whichever dimension of E_q, this inertia is worth 1. In this way, when searching for directions of maximum inertia, these two types of variables are balanced, which is highlighted by one or another of the two expressions of the criterion below.

3.3 Representing Individuals

The dimensions of space $\mathbb{R}^K$ are quantitative variables K_1 and indicators K_2. Its diagonal Euclidian metric is the column weights (1 for the quantitative variables and p_{k_q} for the categories).

The distance between individuals i and l is expressed:

$$d^2(i,l) = \sum_{k \in K_1} (x_{ik} - x_{lk})^2 + \sum_{q \in Q} \sum_{k \in K_q} p_{k_q} \left(\frac{y_{ik_q}}{p_{k_q}} - \frac{y_{lk_q}}{p_{k_q}} \right)^2.$$

The quantitative variables contribute to this distance in exactly the same way as in a PCA on these variables alone (see Section 1.3); the qualitative variables contribute to this distance (up to $1/Q$ coefficient) as they do in the MCA of these variables alone (see Section 2.4). One specific important case is that of the distance between an individual and the cloud's centre of gravity. This centre of gravity lies at the origin O when the variables are centred, as assumed for quantitative variables. For the MCA-coded indicators, accounting for division by p_{k_q}, the mean of column k_q is worth 1 (see Section 2.4).

Finally, we obtain:

$$d^2(i,O) = \sum_{k \in K_1} x_{ik}^2 + \sum_{q \in Q} \sum_{k_q \in K_q} p_{k_q} \left(\frac{y_{ik_q}}{p_{k_q}} - 1 \right)^2 = \sum_{k \in K_1} x_{ik}^2 + \sum_{q \in Q} \frac{1 - p_{q(i)}}{p_{q(i)}},$$

where we denote $q(i)$ the category of variable q possessed by i, and $p_{q(i)}$ the proportion associated with $q(i)$.

It is then necessary to ensure the balance between the influence of the two types of variables in these relationships. It is natural to measure the influence of a variable by its contribution to the inertia of all of the points. The considerations established in $\mathbb{R}^I$ are transposed in $\mathbb{R}^K$ by duality. Particularly, in the subspace of $\mathbb{R}^K$ generated by K_q categories of variable q, the projection of the cloud of individuals has an inertia of $K_q - 1$ distributed isotropically in all directions of this subspace of dimension $K_q - 1$.

As in all factorial analyses we represent:

- The cloud of individuals by its projection on its axes of inertia (we denote $F_s(i)$ the projection of individual i on the axis of rank s).
- The quantitative variables by their correlation coefficient with the factors F_s.
- The categories of qualitative variables by the centres of gravity of the corresponding individuals. We denote $F_s(k_q)$ the coordinate of the projection, on the axis of rank s, of the centre of gravity of the individuals possessing category k of variable q.

3.4 Transition Relations

Here we apply the general formulae for PCA (see Section 1.5.4) to the encoded table as indicated in Section 3.2.

Relationships from $\mathbb{R}^K$ Toward $\mathbb{R}^I$
Let $G_s(k)$ be the coordinate of column k on the axis of rank s.

Case of a quantitative variable:

$$G_s(k) = \frac{1}{\sqrt{\lambda_s}} \sum_i p_i x_{ik} F_s(i) = r(k, F_s).$$

Case of a category k_q of variable q with a relative frequency of p_{k_q}:

$$G_s(k_q) = \frac{1}{\sqrt{\lambda_s}} \frac{1}{p_{k_q}} \sum_i p_i y_{ik_q} F_s(i) = \frac{1}{\sqrt{\lambda_s}} F_s(k_q),$$

where $F_s(k_q)$ is the coordinate, along the axis of rank s, of the centre of gravity of individuals with category (k_q). As in MCA, up to the $1/\sqrt{\lambda_s}$ coefficient, the coordinate of a category as an indicator (that is to say in $\mathbb{R}^I$), is equal to that of the barycentre of the individuals which possess it (in $\mathbb{R}^K$).

Relationship from $\mathbb{R}^I$ Toward $\mathbb{R}^K$
This relationship is fundamental in MCA as it expresses the position of an individual according to the categories which it possesses. It is rarely explicit in PCA but it underlies interpretation. For FAMD, it is expressed:

$$F_s(i) = \frac{1}{\sqrt{\lambda_s}} \sum_{k \in K_1} x_{ik} G_s(k) + \frac{1}{\sqrt{\lambda_s}} \sum_{k_q \in K_2} p_{k_q} \left(\frac{y_{ik_q}}{p_{k_q}} - 1 \right) G_s(k_q).$$

The first member is that of PCA (see Section 1.5.4). It expresses that an individual lies on the side of the variables for which it has an above average value, and opposite variables for which it carries a below average value. The second member is that of MCA, up to the $1/Q$ coefficient (see Section 2.5.3). It can be expressed according to $F_s(k_q)$, thanks to the above equation relating $G_s(k_q)$ with $F_s(k_q)$:

$$\frac{1}{\lambda_s} \sum_{k_q \in K_2} \left(y_{ik_q} - p_{k_q} \right) F_s(k_q) = \frac{1}{\lambda_s} \sum_{k_q \in K_2} y_{ik_q} F_s(k_q).$$

The latter equation expresses that an individual is, up to the λ_s coefficient, at the barycentre of the categories it possesses (with these categories themselves the barycentres of individuals).

Remark
In the transition relation expressing the coordinate of an individual according to those of the categories, the coefficient is:

- $\sqrt{\lambda_s}$ if the categories are represented by the projection of the indicators (in $\mathbb{R}^I$)

- λ_s if the categories are represented by the centres of gravity of the individuals possessing the same category (in $\mathbb{R}^K$)

Finally, an individual is found both on the side of the quantitative variables for which it has a high value, and on the side of the categories it possesses.

3.5 Implementation

The simplest solution is to use an FAMD program such as the `FAMD` function of the FactoMineR package or the AFMD function in UNIWIN Plus. Otherwise, an FAMD can be performed from a regular PCA program. In order to do this, the quantitative variables must first be centred and reduced as we use unstandardised PCA. The qualitative variables appear through their indicators in which y_{ik_q} (= 0 or 1) is divided by $\sqrt{p_{k_q}}$. Dividing by $\sqrt{p_{k_q}}$, rather than p_{k_q} as indicated in Section 3.2, makes it possible to use the metric identity in $\mathbb{R}^K$ (most PCA programs do not permit another metric). Moreover, it is not necessary to centre the data, as this will be done by PCA programs. PCAs such as this directly provide the representations of individuals and quantitative variables. To obtain the representation of the categories' centres of gravity, the qualitative variables need to be introduced as supplementary.

3.6 Example: Biometry of Six Individuals

To illustrate FAMD, we use a small and simple dataset. Six individuals are described by two uncorrelated quantitative variables (height and weight) and one qualitative variable (hair colour, with three categories) connected to the first two. The data are brought together in Table 3.1.

TABLE 3.1
Biometry Data[a]

	Hair Colour	Height	Weight	CR Height	CR Weight
a	Blonde	1	1	−1.464	−1.225
b	Blonde	2	2	−0.878	0.000
c	Brown	3	3	−0.293	1.225
d	Brown	4	3	0.293	1.225
e	Black	5	2	0.878	0.000
f	Black	6	1	1.464	−1.225
Mean		3.5	2	0	0
Standard Deviation		1.708	0.816	1	1

[a] CR: centred-reduced.

TABLE 3.2
Biometry. Relationship Matrix[a]

	Hair Colour	Height	Weight
Hair Colour	2		
Height	0.9143	1	
Weight	0.7500	0	1

[a] Each number, except the diagonal, is the squared correlation coefficient or correlation ratio according to the nature of the variables involved. Diagonal: 1 for the quantitative variables; number of categories minus 1 for the qualitative variable.

Table 3.2, known as the *relationship matrix*, brings together the relationship measurements between the variables taken two by two. These measurements are:

- The squared correlation coefficient in the case of two quantitative variables. The height and weight variables are orthogonal ($r^2 = 0$).
- The squared correlation ratio in the case of a quantitative variable and a qualitative variable. Hair colour is related to both height and weight; it is most closely related to height (.9143 > .7500), which can be read in the data as such: for example, both blondes and those with black hair have identical mean weights but are of very different average height.
- ϕ^2 ($= \chi^2/I$) in the case of two qualitative variables; here, this only concerns the diagonal term corresponding to hair colour. This is the number of categories minus 1.

As expected given the way the data were constructed, the first plane expresses almost all (95.56%) of the inertia (first row of Table 3.3).

Whatever the axis, its inertia is distributed equally between a quantitative variable and the qualitative variable. This example clearly illustrates:

TABLE 3.3
Biometry. FAMD.[a]

	F1	F2	F3	F4
% of inertia	48.91	46.65	3.35	1.10
Eigenvalue	1.96	1.87	0.13	0.04
Hair Colour	0.98	0.93	0.07	0.02
Height	0.98	0	0	0.02
Weight	0	0.93	0.07	0

[a] Inertia decomposed by axis and by variable.

- The effectiveness of FAMD in dealing with the crucial issue of balancing the two types of variables
- The fact that a quantitative variable cannot be closely linked to more than a single axis (as in PCA) whereas a qualitative variable can be closely linked to (number of categories – 1) axes (as in MCA)

Figure 3.2 represents the relative positions of the variables and axes in $\mathbb{R}^I$. Due to the orthogonality between height and weight, the axes are situated on the bisector of the angle formed by a quantitative variable and its projection on the plane generated by the qualitative variable. This figure clearly illustrates the geometric interpretation of the FAMD in the space of functions on I and how the balance is achieved between the two types of variable.

The (absolute) contribution of a variable to the inertia of an axis is also interpreted as a measurement of the relationship between the variable and the axis (squared correlation, ratio or coefficient according to the nature of the variable). These inertias can be represented in the graph known as the relationship square (see Figure 3.3, left) which shows that:

- The first factor results from the variables *height* and *colour*.
- The second factor results from the variables *weight* and *colour*.
- In both cases the contributions are balanced between the two variables.
- In both cases, the factor is closely linked to each of the two variables.

With this general framework defined, interpreting Figures 3.3 and 3.4 is easy in this simple case. The first axis opposes small blondes and tall people with black hair. These two groups of individuals are light and oppose the heavy, brown-haired individuals on the second axis.

3.7 FAMD in FactoMineR

The Biometry data (see Table 3.1) illustrates this analysis.

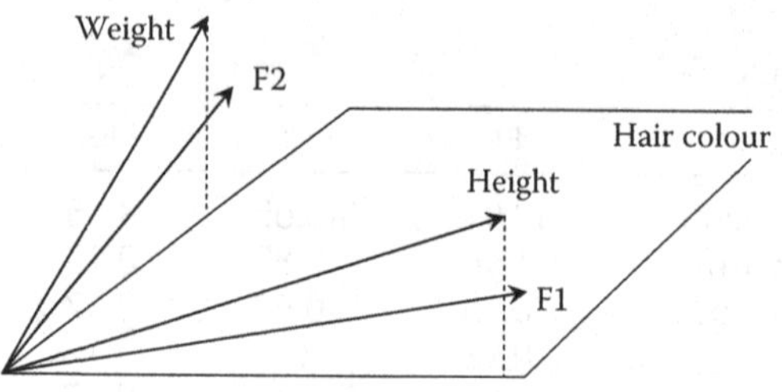

FIGURE 3.2

Biometry. FAMD. Representation of variables and axes in $\mathbb{R}^I$.

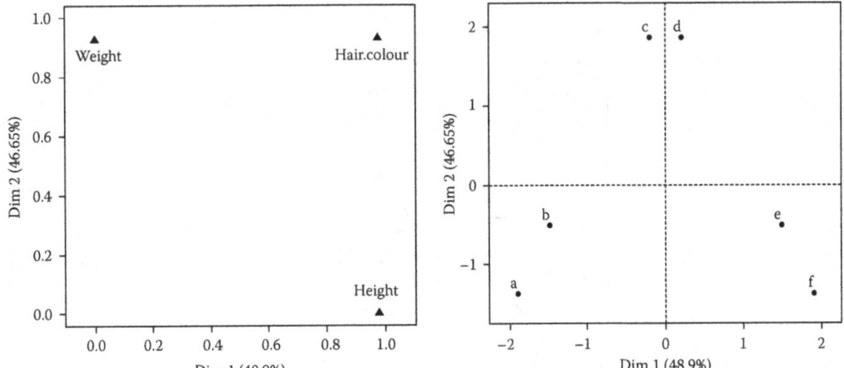

FIGURE 3.3
Biometry. FAMD. Relationship square (left) and representation of individuals (right).

Drop-Down Menu in R Commander

Main Menu (see Figure 3.5)

1. Window in which the active quantitative variables are selected.
2. Window in which the active qualitative variables (factors) are selected.
3. Buttons for opening windows to select supplementary elements.
4. Opens the graph window (see Figure 3.6).
5. The output options make it possible, if need be, to avoid displaying the results concerning individuals (which is vital when there are a large number of them) and to save all of the results in a csv file.
6. A clustering (Ward's method) can be performed from the factorial coordinates.

Graphical Options (see Figure 3.6)

The FAMD function produces three types of graph.

- Graph of individuals (and categories).
  - 1 and 2. Used to select and label the types of elements to be represented (almost) independently.
  - 3. Individuals can be attributed specific colours according to their category of a qualitative variable, either active or supplementary. Here, only one variable is available: colour (of hair).
- Graph of variables (relationship square).
  - 4. The colours of different types of variables can be managed independently.

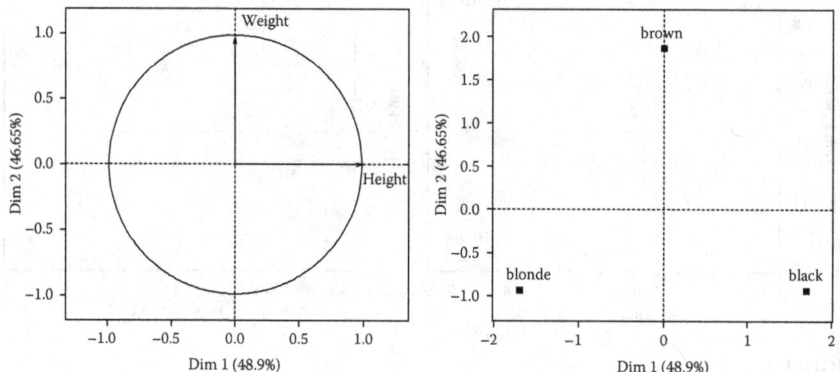

FIGURE 3.4
Biometry. FAMD. Representation of quantitative variables (left) and categories of the quantitative variable (right).

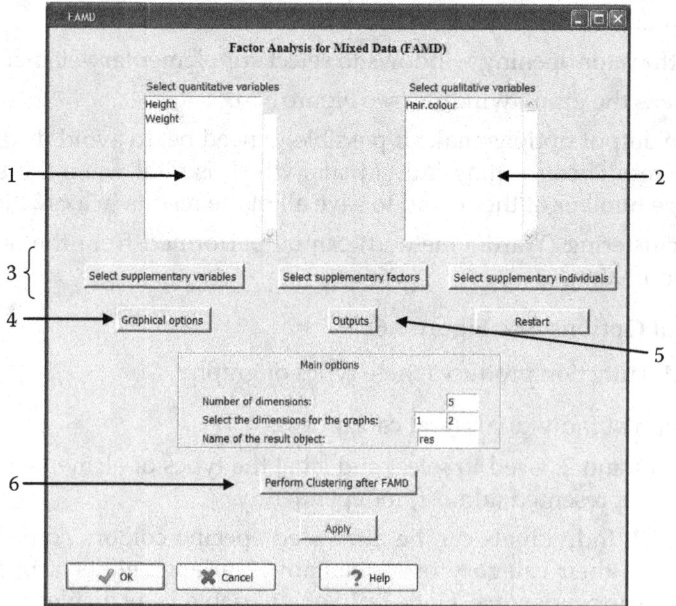

FIGURE 3.5
FAMD main window.

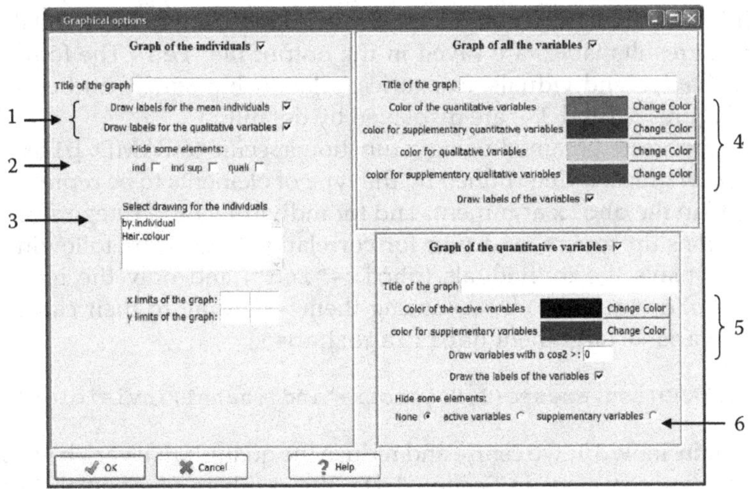

FIGURE 3.6
FAMD Graphical options window.

- Graph of quantitative variables (correlation circle).
  - 5. The colours of different types of quantitative variables can be managed independently.
  - 6. Used to select a type of quantitative variables to be represented.

Examples of Commands

The data are imported and listed for verification.

```
> biometry=read.table("biometry.csv",sep=";",dec=",",
+ header=TRUE, row.names=1)
> biometry
  Hair.colour Height Weight
a      blonde      1      1
b      blonde      2      2
c       brown      3      3
d       brown      4      3
e       black      5      2
f       black      6      1
>
```

The FAMD can be launched by choosing all of the default options.

```
> res=FAMD(biometry)
```

In this case, all of the variables are active, both quantitative and qualitative. All of the results tables are saved in the output file (`res`). The four graphs in this chapter (individuals, categories, relationship square and correlation circle; Figures 3.3 and 3.4) are displayed by default.

The graphs are obtained using a function specific to FAMD: `plot.FAMD`. The type of graph is determined by the type of elements to be represented as specified in the `choix` argument: `ind` for individuals and categories, `group` for relationship square, and `var` for correlation circle. The following command displays the individuals (`choix="ind"`) and only the individuals (`invisible="quali"`) by colouring them according to their category for the first variable (argument `habillage`: `hab=1`).

```
> plot.FAMD(res,axes=c(1,2),choix="ind",hab=1,invisible="quali")
```

To obtain Table 3.1, we centre and reduce the quantitative variables (`scale` function; see comment in Section 1.11); the result is saved in the `tab3_1` matrix.

```
> tab3_1=scale(biometry[,2:3])*sqrt(6/5)
```

Finally, `tab3_1` is horizontally concatenated with the raw data by

```
> cbind(biometry,tab3_1)
```

The relationship matrix (see Table 3.2) is in `res$var$Lg`.

To construct Table 3.3, the eigenvalues and percentages of inertia, which are in the first two columns of `res$eig`, are saved and transposed (`t` function), in the `tab3_3` matrix.

```
> tab3_3=t(res$eig[,c(2,1)])
```

This table is concatenated with `res$var$coord` which contains the inertias of the variables along axes; we select the first four axes, as those which follow have null inertia.

```
> round(rbind(tab3_3[,1:4],res$var$coord[,1:4]),2)
```

4

Weighting Groups of Variables

This chapter is the first devoted to multiple factor analysis (MFA). MFA is applied to tables in which a set of individuals is described by several sets of variables. The key points of this method are dealt with in succession over four chapters. The first aims to balance the groups in an overall analysis by weighting the variables. This characteristic of MFA is vital, in the sense that it is the source of many properties of this method.

4.1 Objectives

In order to keep things simple, we here deal with standardised quantitative variables. As the table to be analysed is an individuals $\times$ variables table, the two main issues associated with it are that of principal component analysis (PCA), that is to say:

1. A representation of the variability of individuals. This is done using the principal dimensions of this variability.
2. A representation of the correlation between variables constructed using synthetic variables (the principal components).

The partition of the variables (into groups) does not appear at this general level. We therefore need to specify the influence of the group structure on the issue at hand. The fundamental idea is that identifying groups among the variables more or less explicitly implies a balance between them (the precise meaning of which still needs to be determined). Indeed, what can be said about an analysis such as this which, for example, might highlight a first factor depending primarily on one group alone? Is this factor not merely the result of an especially high inertia for this group of variables? In this case it is not particularly noteworthy. However, if the inertia of the groups has been balanced (in a sense defined hereafter), users would be able to interpret the impossibility in these data of displaying a direction of high inertia to which the different groups contribute (more or less) equally. Without getting into the technical details, which we do later, we can already specify what we expect from such a balance. We can approach this issue from many different perspectives. As we work within the context of factorial analysis, we focus our attention on the inertia of the clouds studied.

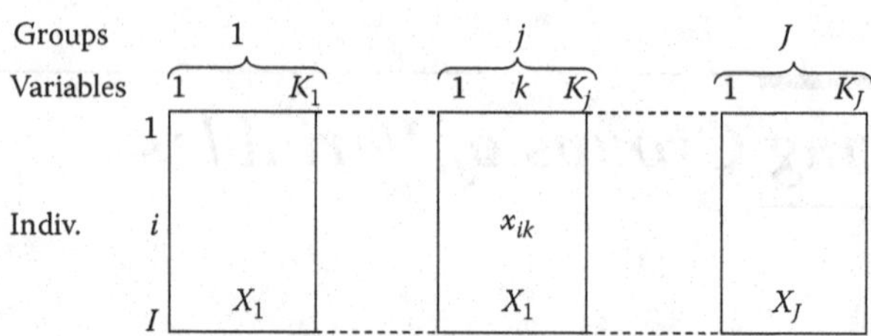

FIGURE 4.1
Table with variables organised into groups.

Notations. We use the notations from PCA (see Figure 4.1): (i, I) for the individuals, (k, K) for the variables and p_i for the weight of individual i ($\sum_i p_i = 1$). These are completed by the notations related to the groups:

- j: Index of the current group
- J: Number of groups (or set of groups)
- K_j: number of variables of group j (or the set of these variables) and $K = \sum_j K_j$.
- X_j: Subtable containing the data from group j alone

The total inertia of the cloud of individuals (identical to that of the cloud of variables) can be expressed, highlighting the role of each group of variables (the data are centred) thus

$$\sum_i p_i d^2(O, i) = \sum_i p_i \sum_k x_{ik}^2 = \sum_j \sum_{k \in K_j} \sum_i p_i x_{ik}^2 = \sum_j \sum_{k \in K_j} \text{Var}[k].$$

In this way, the contribution of group j to the total inertia of the cloud of individuals is equal to the sum of variances of the variables of this group. With standardised variables, this contribution is equal to its number of variables: the influence of a group depends first and foremost on the number of variables. This result has led many authors to suggest standardising (at 1) this contribution (which we obtain by attributing each variable a weight equal to the inverse of the number of variables of the group to which it belongs).

However, aside from the notion of inertia, the above approach does not take into account the factorial analysis approach, that is to say the search for directions of maximum inertia. With this in mind, we express the criteria maximised by the first axis, highlighting the contribution of each group of variables. Reasoning based on the cloud of variables itself is more direct. Denoting H_k^s the projection of variable k on the axis of rank s, the criterion associated with the axis of rank s is expressed:

$$\sum_k (OH_k^s)^2 = \sum_j \sum_{k \in K_j} (OH_k^s)^2 .$$

TABLE 4.1
Data (a) and Correlation Matrix (b)

	$V1$	$V2$	$V3$	$V4$	$V5$		$V1$	$V2$	$V3$	$V4$	$V5$
1	1	1	1	1	1	$V1$	1				
2	1	1	−1	0	0	$V2$	0	1			
3	1	−1	1	0	0	$V3$	0	0	1		
4	1	−1	−1	1	1	$V4$	$\sqrt{2}/2$	0	0	1	
5	−1	1	1	−1	−1	$V5$	$\sqrt{2}/2$	0	0	1	1
6	−1	1	−1	0	0						
7	−1	−1	1	0	0						
8	−1	−1	−1	−1	−1						
			a					b			

The contribution of a group of variables is here the sum of the projected inertias for the variables belonging to this group. We may want to harmonise these contributions directly rather than indirectly as in the previous approach. Indeed, in this new approach, harmonising the total inertias of the groups (for example, at 1) provides an upper limit to the contribution of a group in constructing an axis. However, if the total inertia of a group is equal to one, this limit can only be reached by a one-dimensional group. We may want the balance between the groups to lie within a limit (for contribution) which can be reached by each group (whether this limit is actually reached of course depends on the data).

4.2 Introductory Numerical Example

We illustrate these considerations using an example and reasoning in the variables' space ($\mathbb{R}^I$). Within this space, the variables of group j make up a cloud denoted N_K^j. We analyse the inertia of these clouds N_K^j.

Let us consider eight individuals described by two groups of variables (see Table 4.1). Group 1, $\{V1, V2, V3\}$, is made up of three uncorrelated variables (in this group we recognise the complete experimental design for three factors with two levels but these factors are here considered quantitative). Group 2, $\{V4, V5\}$, is made up of two identical variables, correlated with only one variable from group 1 ($V1$). Table 4.2A brings together the results of the PCA of Table 4.1 which deal with the variables and the first axis. This first axis is mainly due to group 2: the relative contribution of its variables to axis 1 is 72.36%. Therefore, in this example (which, it is true, is custom designed), the smallest group has the greatest influence on the first axis.

In this way we illustrate the shortcoming of the notion of total inertia in defining the influence of a group: the distribution of this inertia must also be

TABLE 4.2
Coordinates and Contributions of the Variables[a]

A	Coordinate	Contribution	Contribution (%)
$V1$	0.85	0.72	27.64
$V2$	0	0	0
$V3$	0.00	0.00	0
$V4$	0.97	0.95	36.18
$V5$	0.97	0.95	36.18
Total		2.62	100
B	**Coordinate**	**Contribution**	**Contribution (%)**
$V1$	0.811	0.66	18.38
$V2$	0	0	0
$V3$	0	0	0
$V4$	0.99	0.97	27.21
$V5$	0.99	0.97	27.21
$V6$	0.99	0.97	27.21
Total		3.581	100

[a] For the first axis of the PCA of Table 4.1 A or Table 4.1 in which we added variable $V6$ identical to $V4$ and $V5$ (B).

taken into account. In the example, group 2 has a total inertia of 2, concentrated in only one direction. Group 1 has a higher total inertia (3) but this inertia is equally distributed in a three-dimensional subspace. Factorial analysis, which looks for directions of high inertia is, by definition, sensitive to the distribution of inertia within the groups; this is why we must account for this distribution when looking for a balance between groups. Here we find the idea mentioned in the initial issue of taking into account the statistical method when balancing the groups.

By balancing the total inertia of each group, we reinforce the influence of group 2 on the first axis: the contribution of this group to this axis then increases to 81.62% (see Table 4.2B; in this example, weighting by the total inertia (3) is obtained by introducing variable $V6$, identical to $V4$ and $V5$ and using a usual PCA program). Although this example was chosen for its simplicity, it is not unrealistic: just imagine any situation in which one group is reduced to a single variable when the other has several. In such cases, weighting by total inertia would lead to a first axis almost equal to the variable of the group which has only one.

4.3 Weighting Variables in MFA

Taking into account the distribution of inertia in such a way that it can be applied no matter how many variables there are, means considering the

TABLE 4.3
Coordinates and Contributions of the Variables[a]

Label	Weight	Coordinate	Contribution	Contribution (%)
$V1$	1	0.924	0.854	50
$V2$	1	0	0	0
$V3$	1	0	0	0
$V4$	0.5	0.924	0.427	25
$V5$	0.5	0.924	0.427	25
Total			1.707	100

[a] For the first axis of MFA applied to Table 4.1.

principal direction of inertia alone (every group has one). For a given group j, this means attributing a weight to each variable with these two features:

1. This weight is the same for each variable in the group; in this way we do not distort the distribution of inertia within each group j.
2. Maximum axial inertia is equal to 1 (this inertia is the first eigenvalue of the separate PCA for group j accounting for the weights). In order to do this, we attribute each variable in group j the weight $1/\lambda_1^j$ where λ_1^j is the first eigenvalue of the separate PCA for group j. The PCA of the full table with these weights is the core of the MFA.

In the example, this means attributing a weight of 1 to each variable in group 1 and a weight of 1–2 to each variable in group 2. The coordinates and contributions of the variables associated with the first axis of the MFA are brought together in Table 4.3. In this analysis, the groups' contributions to the first axis (that is to say the sums of the contributions of the variables of a given group) are identical. Weighting has worked perfectly.

Figure 4.2 geometrically represents the relative positions (in $\mathbb{R}^I$) of the first factor of the three analyses mentioned and the variables which contributed to their construction. The first factor of usual PCA is more 'attracted' by group 2 ($V4$, $V5$) than by group 1. This attraction is reinforced with the weighting balancing the total inertias and is cancelled in MFA.

In practice, the influence of weighting on the construction of the first axis in MFA is highly changeable depending on the data; this influence increases along with the difference between the first eigenvalues of the PCAs. But in any case, standardising the maximum axial inertia of each group induces properties which are highly valuable in the interpretation phase.

What about the subsequent axes in MFA? Table 4.4 brings together the inertias of the two groups of variables from Table 4.1, in MFA and in the separate PCAs.

It shows that axes 2 and 3 of the MFA are generated by group 1 alone. This is not contradictory to the desire to balance the groups: a multidimensional

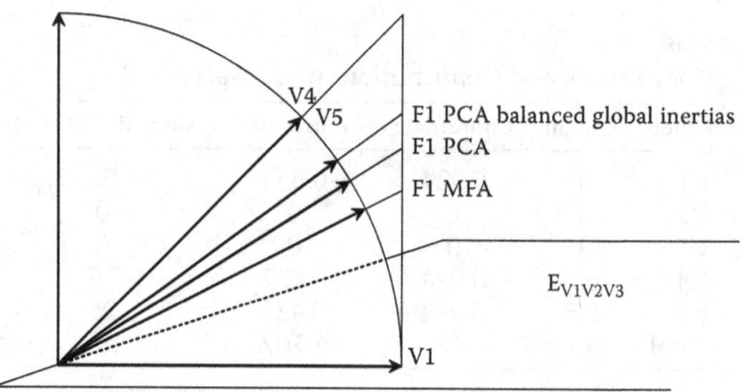

FIGURE 4.2
Representation in $\mathbb{R}^I$ of the variables from Table 4.1 and the first standardised factors of the analyses. E_{V1V2V3}: subspace generated by the three variables of group 1. As the graph suggests, $V1$ is collinear to the projection of $V4$ on E_{V1V2V3}.

group (group 1) is expected to contribute strongly to more dimensions than a one-dimensional group.

This idea also appears in multiple correspondances analysis (MCA). A qualitative variable with many categories has great inertia (equal to the number of categories minus 1), but this inertia is distributed isotropically in a high-dimensional subspace (the number of dimensions is equal to the number of categories minus 1). A variable like this alone cannot generate the first axis, but it can play a role in several axes. This relationship (between MFA and MCA) is not mere chance: this property of MCA inspired the weighting in MFA.

Weighting by maximum axial inertia can be applied directly to groups of unstandardised variables. Therefore, it is possible to account simultaneously for groups made up of standardised or unstandardised variables. The Orange Juice dataset provides a good example of this: some sensory scientists analyse products × descriptors tables using unstandardised PCA (to attribute more importance to the descriptors which vary greatly from one product to another). The user who adopts this point of view is led to perform an MFA in

TABLE 4.4
Total Inertia Decomposition in MFA and in the Separate PCAs

	Total Inertia	Axis 1	Axis 2	Axis 3	Axis 4
Separate Analyses					
Group 1	3	1	1	1	0
Group 2	2	2	0	0	0
MFA					
Total	4	1.71	1	1	0.29
Group 1	3	0.85	1	1	0.15
Group 2	1	0.85	0	0	0.15

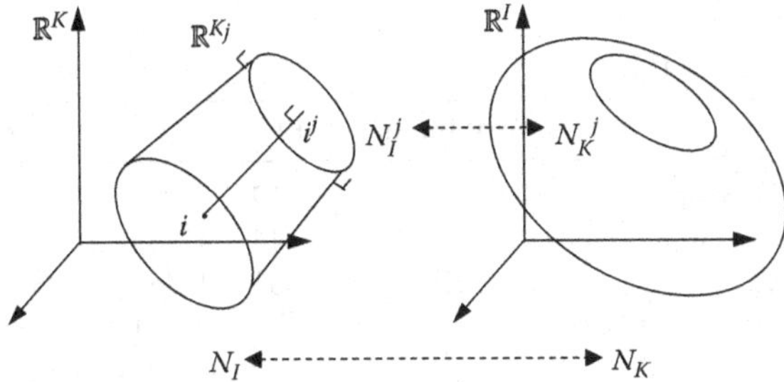

FIGURE 4.3
Duality between clouds of individuals and variables; clouds complete (N_I, N_K) or restricted to group of variables j (N_I^j, N_K^j).

which the sensory group is composed of unstandardised variables and the chemical group of standardised variables. By weighting the variables in such a way as to standardise (to 1) the maximum axial inertia of each group, these two types of variables can be simultaneously introduced as active within one analysis.

In the individuals' space $\mathbb{R}^K$, the attribution of weights to the variables induces a specific metric: in calculating a distance, each variable of group j intervenes with a weight of $1/\lambda_1^j$. Thus, by highlighting part of group j within the distance between individual i and the origin O:

$$d^2(O,i) = \sum_j \frac{1}{\lambda_1^j} \sum_{k \in K_j} x_{ik}^2.$$

The part of group j in $d^2(O,i)$ is interpreted geometrically as the squared distance between the origin and the projection of i (denoted i^j) on the subspace generated by variables of group j (denoted $\mathbb{R}^{K_j}$). Individuals' space $\mathbb{R}^K$ is the direct sum of (sub)spaces $\mathbb{R}^{K_j}$ and we can write:

$$d^2(O,i) = \sum_j d^2(O,i^j).$$

To interpret in $\mathbb{R}^K$ the property of weighting the variables introduced in $\mathbb{R}^I$, it is important to consider all points i^j associated with group j: $\{i^j, i = 1, I\}$ (see Figure 4.3). They make up the cloud denoted N_I^j, the projection of cloud N_I on the subspace generated by the variables of group j. This cloud is linked to cloud N_K^j (of the variables of group j in $\mathbb{R}^I$) by relationships of duality. Thus, these two clouds have:

- The same total inertia
- The same inertia projected on the principal dimension of inertia of rank s, particularly the first of these, λ_1^j

TABLE 4.5
Orange Juice Total Inertia[a]

		Total Inertia	F1	F2	F1 (%)	F2 (%)
1	PCA Chemical	8	6.212	1.1	77.66	13.74
2	PCA Sensory	7	4.744	1.333	67.77	19.05
3	PCA All	15	9.801	1.886	65.34	12.57
4	Chemical Group	8	5.727	0.691	58.43	36.65
5	Sensory Group	7	4.075	1.194	41.57	63.35
6	MFA	2.763	1.785	0.365	64.6	13.21
7	Chemical Group	1.286	0.891	0.099	49.92	27
8	Sensory Group	1.476	0.894	0.266	50.08	73

[a] In four analyses, decomposed onto the first two axes and by group when appropriate.

The weighting of the variables introduced in $\mathbb{R}^I$ induces the following property in $\mathbb{R}^K$: the projection of the cloud of individuals on the subspace generated by the variables of group j has a maximum axial inertia of 1. Thus, in $\mathbb{R}^K$, the weighting of the variables is represented by a balance between the groups of variables in the shape of cloud N_I:

- Not in the overall distances between individuals (which would mean a balance between the total group inertias)
- But rather in the absence of a very high direction of inertia which would be the prerogative of a single group

In the individuals' space, duality induces the effect of weighting described in $\mathbb{R}^I$: in comparison with a one-dimensional group of variables, a multidimensional group will not be favoured in the construction of the first axis of inertia, but will influence a greater number of axes.

4.4 Application to the Six Orange Juices

We apply the different methods to the Orange Juice data (see Section 1.10). Table 4.5 summarises the principal decompositions of inertia.

Rows 1 and 2. The two groups present a first predominant axis. This predominance is greater in the *chemical* group (77.66% versus 67.77), which, along with a higher number of variables (8 versus 7), leads to a greater eigenvalue (6.212 versus 4.744).

Rows 3, 4 and 5. This higher maximum axial inertia in group 1 induces, in the usual PCA applied to these two groups, a greater contribution of group 1 to the first axis (58.43% versus 41.57). The first eigenvalue of this PCA is extremely

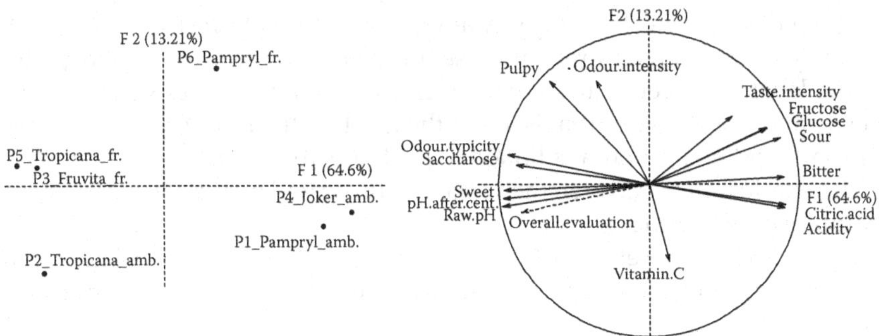

FIGURE 4.4
Orange Juice. MFA. Representation of the individuals and variables on the first plane.

high (9.801) compared to the first eigenvalues of the separate PCAs; it is close to the maximum possible (10.956) which would have been reached if the first principal components (F1) of the separate PCAs had been collinear.

Rows 6, 7 and 8. In MFA, the contributions to the first axis of each of the two groups of variables are almost identical (49.92 and 50.08): here weighting has played its role perfectly. The first eigenvalue can be considered rather high: 1.785, a value close to the maximum possible 2 (which would have been reached if the first principal components ($F1$) of the separate PCAs had been collinear) than to the minimum 1 (which corresponds to cases where each variable of group 1 is uncorrelated with each variable of group 2).

In this case, reading the representation of individuals and variables is the same as for PCA. The first plane expresses a high percentage of inertia (77.81%) justifying our choice to limit ourselves to this plane, at least in this methodological presentation.

The first axis (see Figure 4.4) separates juices P1, P4 and P6 from the other three. According to the representation of variables, these three juices are characterised, relative to the other three:

- From a sensory perspective by a sour, bitter and unsweet taste, and an atypical odour
- From a chemical perspective by a low pH, a high glucose and fructose content and a low sucrose content

The other three juices, P2, P3 and P5, present the opposite characteristics. This axis could be summarised by the opposition *soft juices* ⇔ *hard juices*; this opposition coincides with the origin of the juices, with the 'soft' juices from Florida.

The second axis does not seem so easy to interpret. However, the second bisector practically coincides with the *pulpy* variable. On the plane of individuals, this second bisector corresponds exactly to the opposition between the *refrigerated* (P3, P4 and P6) and *ambient* juices. This relationship between the type of juice and the pulpy characteristic seems to result from a choice made by juice manufacturers because this characteristic cannot be explained by the level of pasteurisation. However, the relationship between odour intensity and this bisector (that is to say the fact that the refrigerated juices have a stronger odour) is perhaps linked to pasteurisation level.

The relationships between the variables illustrated by the correlation circles requires some comment.

The relationship between the three sugars and the pH is related to the hydrolysis of sucrose (in glucose and fructose) favoured in an acidic environment.

We cannot interpret the position of *sweet* from the opposition between the sugars inasmuch as each sugar leads to a sensation of sweetness. This is why we chose to add them together. We therefore obtain a *total sugars* variable which, introduced as supplementary, shows a correlation coefficient of 0.1182 with the first principal component. Therefore, in these data, this variable is not related to the *hard* ⇔ *soft* opposition. Amongst other things, its correlation coefficient with the characteristic *sweet* is 0.1856; we must therefore search for the origin of this sweetness elsewhere.

From a sensory point of view, the first axis evokes the notion of *taste balance* (we perceive the tastes simultaneously and therefore these perceptions are not independent). This context suggests that, in these data, a strong sensation of sweetness is more closely related to an absence of sourness than to a high amount of sugar. The total quantity of sugar also strongly influences this interpretation as it varies from 81 g/L to 101 g/L, a variation which can be considered weak due to the average content (and which is therefore hardly perceptible). As these levels are high, it can be said that, in these data, the 'hardness' (sourness + bitterness) of certain juices masks the influence of sugar.

The representation of individuals and variables is therefore interpreted in the same way as in PCA. The only difference, but only compared to standardised PCA, is that the presence of weights for the variables implies that it is not possible to interpret the correlation circle in terms of the contribution of the variables. Users who want to visualise these contributions will need to construct a supplementary graph.

4.5 Relationships with Separate Analyses

We have already insisted on the need to connect the inertias from the MFA and those in the separate PCAs (see Tables 4.4 and 4.5). It is also useful to connect the factors of the MFA with those of the separate PCAs (also known as *partial axes*), both to understand better the effects of weighting and to enrich interpretation of the analyses. In order to do this, the latter are projected as supplementary variables (see Figure 4.5).

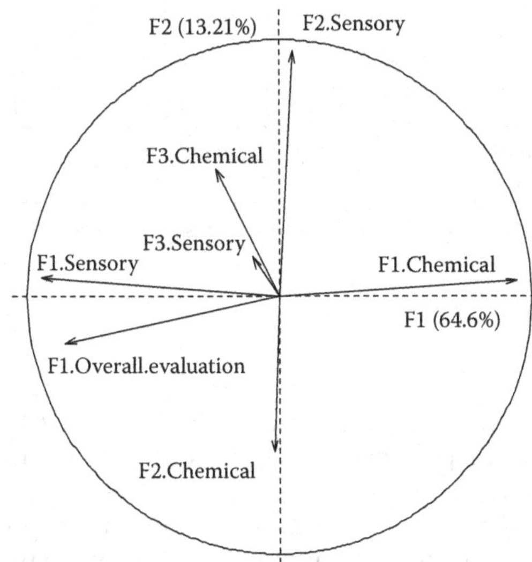

FIGURE 4.5
Orange Juice. MFA. Representation of the first factors of the separate PCAs.

The first factor of the MFA is closely related to the first factor of each group: the principal dimension of sensory variability is correlated with the first dimension of chemical variability and the first factor of the MFA is some kind of compromise between these two directions. The value of this first factor is thus reinforced.

The second factor is closely correlated with the second principal sensory component, which suits the high dimensionality of this group compared with that of the chemical group: MFA weighting standardises the first direction of inertia alone. Finally, the first plane of the MFA is very similar to the first plane of the separate PCA of the sensory group.

Another aspect of the confrontation between the factors of the MFA on the one hand and the principal components of the separate PCAs lies in the representation quality of the variables of each group (that is to say N_K^j) in each analysis. We thus calculate the (projected inertia)/(total inertia) ratio for each N_K^j in the MFA. Therefore, for group j and axis s (of the MFA):

$$\left[\sum_{k\in K_j} \left(OH_k^s\right)^2\right] \frac{1}{\sum_t \lambda_t^j}.$$

These values are to be confronted with the percentages of inertia of the separate PCAs. In the example (see Table 4.6), on the first plane of the MFA,

TABLE 4.6
Orange Juice. MFA[a]

	*F*1	*F*2	Plane (1,2)
Chemical	69.21	7.66	76.86
Sensory	60.58	18.06	78.64

[a] Representation quality of the groups of variables on the first plane (in $\mathbb{R}^I$).

the representation quality of the sensory variables is 78.64%. This value can be compared with the percentage of inertia of the first plane of the PCA for the sensory group (86.81%; see Table 4.5); the difference between the two can be considered as the price of going from an optimal representation of the single sensory group to an optimal representation of the two groups. In this example, we can consider this difference to be small.

It must be noted that we do not change the analysis by replacing a group of variables by all of the principal components of the separate analyses (either unstandardised, or standardised and attributed a weight equal to their inertia taking into account the inertia of the weighting of the MFA). These principal components can thereby be considered as active variables. In particular, their projected inertia can be interpreted as a contribution. These contributions, divided by the total inertia of the axes of MFA, are given in Table 4.7.

The contributions complement the correlation circle (see Figure 4.5) as it does not feature the weights. In this example, these contributions clearly show the essential role of the second component of the sensory group for the inertia of axis 2 of the MFA.

TABLE 4.7
Orange Juice. MFA[a]

Factor on *I*	Weight	*F*1 MFA	*F*2 MFA
$F1$ PCA Chemical	1.000	49.632	1.089
$F2$ PCA Chemical	0.177	0.004	17.683
$F3$ PCA Chemical	0.059	0.205	3.967
$F1$ PCA Sensory	1.000	49.890	1.530
$F2$ PCA Sensory	0.281	0.047	70.180
$F3$ PCA Sensory	0.173	0.106	1.136

[a] Contribution (in %) of the factors of the separate analyses. Weight: eigenvalues, of the separate PCAs, divided by the first of them.

4.6 Conclusion

In this example, the effect of weighting as a balancing factor between the groups is not excessive as the first eigenvalue varies only slightly from one group to another. Consequently, for this dataset, PCA and MFA generate similar axes. But when the first eigenvalue differs greatly from one group to another, this weighting plays a vital role.

However, setting the maximum axial inertia to 1 makes it easier to read the indicators of inertia (see rows 6, 7 and 8 of Table 4.5), which are vital later on. But especially, the advantage of the MFA lies in all of the aspects and are dealt with in the following chapters. In this case, from here on in, the simplified representation of the factors of the separate analysis illustrates the advantage of accounting for structuring the variables into groups. For example, saying that the principal dimension that we present (that of the MFA) is very close to the principal dimension of each group increases the usefulness of the results.

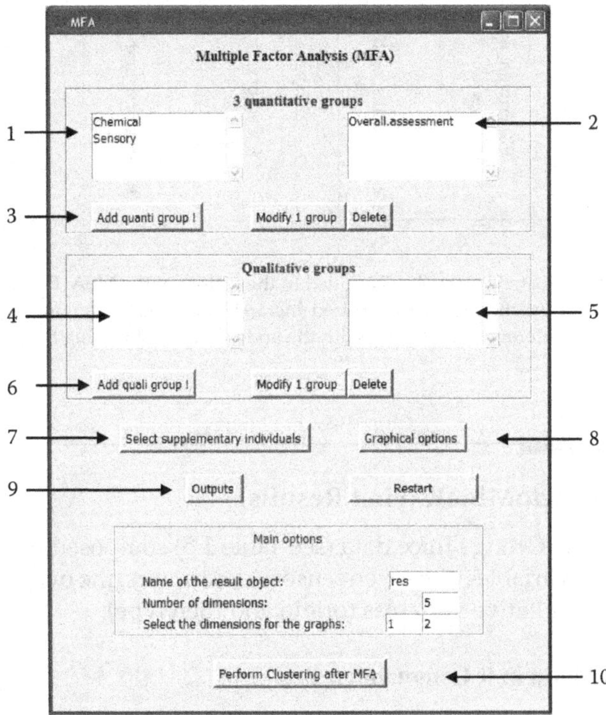

FIGURE 4.6
Main MFA window in **FactoMineR** (via **R Commander**).

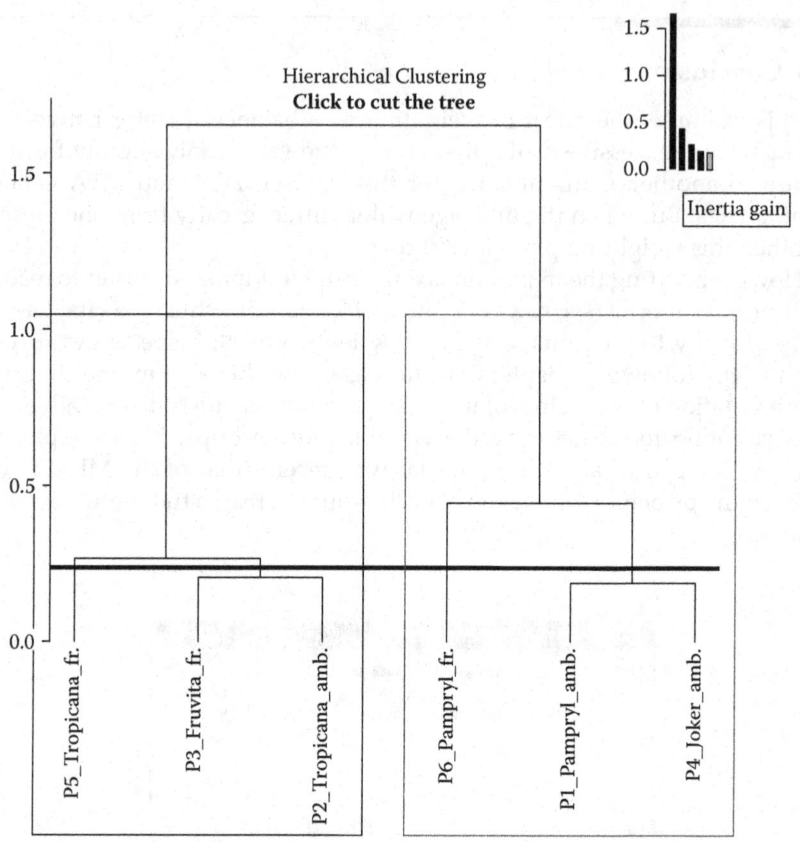

FIGURE 4.7
Orange Juice. Tree created by the AHC applied to the factors of the MFA. Partitioning into two classes was chosen by clicking at the desired hierarchical level. In the insert (top right), the diagram of inertia differences associated with the nodes helps in choosing the cutting level.

4.7 MFA in FactoMineR (First Results)

Here we use the Orange Juice data (see Table 1.5) composed of, in order, the eight chemical variables, the seven sensory descriptors, the overall evaluation and the two qualitative variables (origin and then type).

Drop-Down Menu in R Commander

1. Active quantitative groups. Here, the two groups (chemical and sensory) have already been defined (see point 3 below).
2. Supplementary quantitative groups. Here, the overall evaluation has been introduced as a supplementary group (with one single variable).

3. The "add quanti group" button opens the window in Figure 4.9.
4, 5, 6. Define qualitative groups (see Chapter 8 on qualitative variables).
7. Brings up the list of individuals to select supplementary individuals. By default, all the individuals are active.
8. Opens the graphical options window, Figure 4.10.
9. By default, the program lists all the results tables. This window is used to list just some of them and, if desired, to export them in a csv file.
10. It is possible to follow this up with an ascending hierarchical classification (AHC; Ward's method) from the factorial coordinates (in the same way as following any factorial analysis). The MFA thus acts as preprocessing, first by balancing the groups of variables and then by ranking the factors in order to select the first ones. There are two important characteristics of the AHC in FactoMineR (`HCPC` function):
    1. A representation of the hierarchical tree which makes it possible to choose a cutting level (and therefore a partition of the individuals) interactively (by clicking on an aggregation level; see Figure 4.7).
    2. A representation of the hierarchical tree associated with the factorial representation (see Figure 4.8).

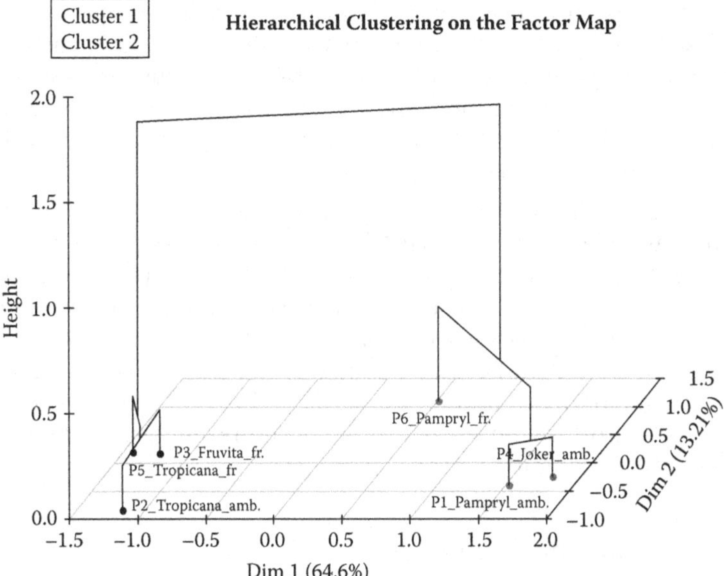

FIGURE 4.8
Orange Juice. Tree from Figure 4.7 associated with the first plane of the MFA.

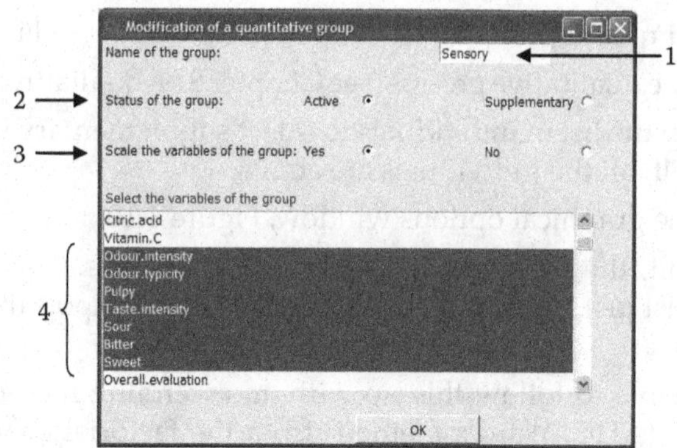

FIGURE 4.9
Window for defining the groups.

Defining the Groups (See Figure 4.9)

1. A name can be attributed to each group. This name is used in the tables and graphs.
2. The group's status (active/supplementary) is defined for each group independently (and thus independently of the order of the variables in the file).
3. (For groups of quantitative variables only) The choice of whether to reduce the data is made for each group independently. It is therefore possible to introduce, within the same analysis, groups which are reduced and groups which are not. Weighting makes this statistically possible. A concrete example would be to introduce the same data twice, once reducing them and the other time without doing so. In this case, MFA is used to compare standardised and unstandardised PCA on the same data.
4. The window lists all of the quantitative variables. The group's variables are selected from this list (they are not necessarily consecutive in the file). When one group is constructed, its variables remain in the list and are thus eligible for another group. It is therefore possible to introduce one variable in several groups, which is useful in methodological studies such as that mentioned in 3 (comparison between standardised and unstandardized PCA) or when we want to compare groups with slightly different compositions (having several variables in common).

Graphical Options (See Figure 4.10)

1 and 2. The graph of individuals (see Figure 4.4, left) can contain partial individuals (Chapter 5) as well as the categories of qualitative

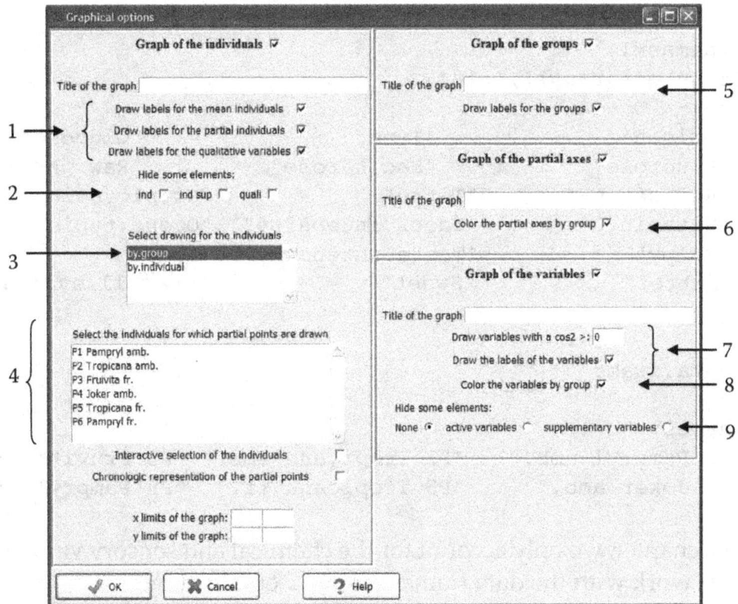

FIGURE 4.10
Graphical options window.

variables (Chapter 8). The presence and labelling of these three types of elements are managed independently.

3. Each individual can be coloured according to a qualitative variable, if appropriate. The 'By group' and 'By individual' options mainly concern partial individuals (Chapter 5).
4. See Chapter 5.
5. See Chapter 7.
6. The partial axes are the axes of separate analyses (see Figure 4.5). By default, the first five axes are represented. Different colours can be attributed to the axes according to groups.
7, 8, 9. The graph of quantitative variables (see Figure 4.4, right) contains the active and/or illustrative quantitative variables (9). It is generally helpful to colour them according to the group to which they belong (8). When there are a large number of variables, it is helpful (7) either not to label them in order to focus on the general shape of the cloud, or to select those which are best represented.

Command Examples

The data are found in the file `OrangeGB.csv` which is imported and for which we check the first 18 columns by

```
> orange=read.table("OrangeGB.csv",sep=";",dec=",",header=TRUE,
+ row.names=1)
> attributes(orange[,1:18])
$names
 [1] "Origin"              "Type"                "Glucose"
 [4] "Fructose"            "Saccharose"          "Raw.pH"
 [7] "Refined.pH"       "Titre"             "Citric.acid"
[10] "Vitamin.C"        "Odour.intensity"   "Odour.typicity"
[13] "Pulpy"            "Taste.intensity"   "Sour"
[16] "Bitter"           "Sweet"             "Overall.evaluation"

$class
[1] "data.frame"

$row.names
[1] "P1 Pampryl amb."    "P2 Tropicana amb." "P3 Fruvita fr."
[4] "P4 Joker amb."      "P5 Tropicana fr."  "P6 Pampryl fr."
```

In this chapter, we only account for the chemical and sensory variables. We therefore work with the data.frame `orange1` obtained by

```
> orange1=orange[,3:18]
```

To perform the MFA on a data.frame, the variables of a given group must be adjacent in the file. This is the case here. We comment hereafter the following command performing MFA:

```
> ResAFM=MFA(orange1,group=c(8,7,1),type=c("s","s","s"),
+ name.group=c("Chemical","Sensory","Overall evaluation"),
+ num.group.sup=3)
```

`group=c(8,7,1)`: there are three groups made up of the first eight variables, the following seven and the last, respectively. The groups are numbered in the order of their variables within the data.

`type=c("s","s","s")`: For all three groups the variables need to be centred and reduced (`"s"` for `scale`; or, `"c"` simply to centre).

`num.group.sup=3`: The third group is supplementary. By default, all the groups are active.

By default this command creates all of the graphs for the first two axes, particularly that of the individuals (see Figure 4.4, left), that of the active quantitative variables (see Figure 4.4, right) and that of the partial axes (see Figure 5.4 to which we add the supplementary variable *overall evaluation*).

The `ResMFA` list contains all of the results tables. These tables are exported into a csv file as follows.

```
> write.infile(ResMFA,file="ResMFA.csv")
```

The graphs are obtained using the `plot.MFA` function, devoted to MFA. The default options are often sufficient, at least at first. The representation of the quantitative variables for all groups on the first plane is obtained by

```
> plot.MFA(ResMFA,choix="var")
```

The following command:

```
> plot.MFA(ResMFA,axes=c(3,4),choix="var",
+ invisible="quanti.sup",hab="group")
```

represents on the plane (3,4) the only active variables (`invisible="quanti.sup"`) coloured (argument `habillage` shortened to `hab`) according to the group to which they belong (`hab="group"`).

Other objects can be represented using the argument `choix`, equal to `ind` for the individuals and to `axes` for the partial axes.

To display the eigenvalues in a bar plot (six individuals and thus at most five nonzero eigenvalues):

```
> barplot(ResMFA$eig[,1],names=1:5)
```

Table 4.5

Users often want to generate a specific table bringing together scattered results. This is the case for Table 4.5, which is given below as an example.

We initialise a matrix (8, 5) named `tab4_5`.

```
> tab4_5=matrix(nrow=8,ncol=5)
```

Row and column names are chosen:

```
> row.names(tab4_5)=c("PCA Chemical","PCA Sensory","PCA All",
+ "Group Chemical","Group Sensory","MFA",
+ "Group Chemical","Group Sensory")
> colnames(tab4_5)=c("Global inertia","F1","F2","F1%","F2%")
```

The first five elements of column 1 are specified:

```
> tab4_5[1:5,1]=c(8,7,15,8,7)
```

Rows (3, 4, 5) of the PCA on the 15 variables require this PCA to be performed.

```
> resPCA=PCA(orange[,1:15])
```

Row 3. The eigenvalues of the PCA are in `resPCA$eig`:

```
> tab4_5[3,2:5]=c(t(resPCA$eig[1:2,1]),t(resPCA$eig[1:2,2]))
```

Rows 4 and 5, columns 4 and 5. The contributions of the variables are in `resPCA$var$contrib`; they must be added together by group.

```
> tab4_5[4,4:5]=apply(resPCA$var$contrib[1:8,1:2],MARGIN=2,
+ FUN=sum)
> tab4_5[5,4:5]=apply(resPCA$var$contrib[9:15,1:2],MARGIN=2,
+ FUN=sum)
```

Rows 4 and 5, columns 2 and 3. The inertia is determined by multiplying the percentage by the eigenvalue.

```
> tab4_5[4:5,2]=tab4_5[4:5,4]*resPCA$eig[1,1]/100
> tab4_5[4:5.3]=tab4_5[4:5.5]*resPCA$eig[2.1]/100
```

The separate PCAs of each group (rows 1 and 2) are listed via the MFA in `ResMFA$separate.analyses`, with the eigenvalues given in `$eig`:

```
> tab4_5[1,2:5]=c(t(ResMFA$separate.analyses$Chemical$eig
+ [1:2,1]), t(ResMFA$separate.analyses$Chemical$eig[1:2,2]))
> tab4_5[2,2:5]=c(t(ResMFA$separate.analyses$Sensory$eig
+ [1:2,1]), t(ResMFA$separate.analyses$Sensory$eig[1:2,2]))
```

Row 6. The eigenvalues of the MFA are in `ResMFA$eig`:

```
> tab4_5[6,2:5]=c(t(ResMFA$eig[1:2,1]),t(ResMFA$eig[1:2,2]))
```

Rows 6, 7, 8, column 1. The overall inertia of a group in the MFA is a direct result of the number of variables (the variables are reduced) and weighting (by the first eigenvalue):

```
> tab4_5[7:8,1]=tab4_5[1:2,1]/tab4_5[1:2,2]
> tab4_5[6,1]=tab4_5[7,1]+tab4_5[8,1]
```

Rows 7 and 8. The inertias of the variables cumulated by group are in `ResMFA$group`; the raw inertias are in `coord` (this term is explained in Chapter 7) and the percentages in `contrib`:

```
> tab4_5[7:8,2:3]=c(t(ResMFA$group$coord[,1]),
+ t(ResMFA$group$coord [,2]))
> tab4_5[7:8,4:5]=c(t(ResMFA$group$contrib[,1]),
+ t(ResMFA$group$contrib[,2]))
```

For the listing, we reduce the number of decimal places.

```
> tab4_5[,2:3]=round(tab4_5[,2:3],3)
> tab4_5[,4:5]=round(tab4_5[,4:5],2)
> tab4_5[1:5,1]=round(tab4_5[1:6,1],0)
> tab4_5[6:8,1]=round(tab4_5[7:8,1],3)
> tab4_5
```

The display in R uses the same number of decimal places in each column, which here is three for column 1. However, the different number of decimal places requested for this column is respected in the exportation in the .csv file.

Simplified Output

The MFA produces many results tables. The `summary.MFA` function summarises the main tables. The following command generates these tables in the `SorMFA` text file.

```
summary(ResMFA,nbelements=Inf,file="SorMFA")
```

This function exists for all factorial analyses in FactoMineR.

5

Comparing Clouds of Partial Individuals

Much multiple factor analysis (MFA) use involves comparing subtables, each made up of a group of variables. One aspect of this comparison was described in the previous chapter that of the factors of separate analyses. Here we deal with the aspect that undoubtedly contributes the most to the richness of MFA results: comparing clouds of individuals each associated with one of the groups of variables.

5.1 Objectives

A row of a data table in which we only consider the data relative to a single group of variables is known as a *partial individual*. Thus, partial individual i corresponding to group j, denoted i^j, is made up of all of the values of $\{x_{ik}; k \in K_j\}$. The set of partial individuals $\{i^j, i \in I\}$ associated with group j makes up partial cloud N_I^j. This cloud of individuals is analysed when conducting a principal component analysis (PCA) of the data of the single group j. It evolves in the space generated by the variables of group j alone: $\mathbb{R}^{K_j}$. In MFA we therefore have to consider $J+1$ clouds of individuals: the J partial clouds to which we add cloud N_I, a cloud we can consider 'overall' (as opposed to partial) evoking the fact that it is associated with the whole set of (active) data. In practice, however, we designate it a *mean cloud*, in reference to one of its geometric properties.

The comparison of partial clouds is the geometric translation of the question of whether two individuals i and l which are similar in terms of group j are also similar in terms of group g.

Thus, when considering the groups of sensory and chemical variables in the example of the six orange juices, we might ask ourselves the following questions:

- Overall, do two orange juices which are similar from a chemical point of view also have the same sensory profile?
- Are there any juices which have a 'mean' chemical profile and an unusual sensory profile? Could such a discrepancy arise from, for example, a chemical characteristic which was not measured, an interaction between chemical characteristics (on perception) or some

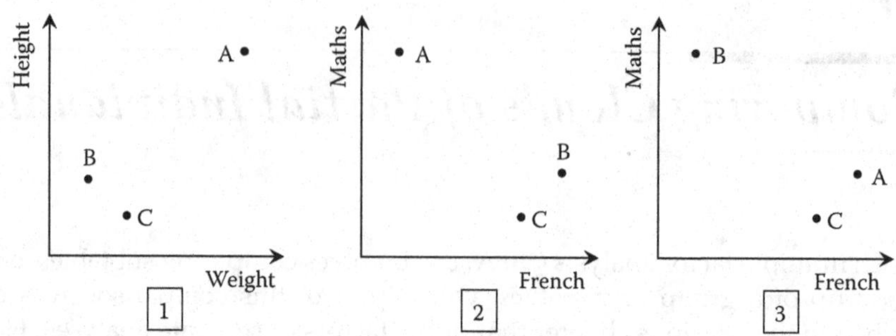

FIGURE 5.1
Three individuals (A, B, C) according to two groups of variables: biometry (1) and grades (two variations: 2 and 3).

other factor? Users expect statistics to identify such discrepancies so that they might then be interpreted.

We can illustrate our expectations when comparing (the shape of) partial clouds by using a small example of data chosen for its simplicity. For three students (A, B and C), we have biometric measurements (height and weight) and exam grades (French and maths). As both groups are two-dimensional, it is possible to represent partial clouds graphically, with the visual examination acting as a statistical analysis (see Figure 5.1).

In the first case (graphs 1 and 2), the graphs show a similar shape for the two clouds. In each, students B and C are very similar and A is particular. According to the graphs, A is both tall and heavy, and has a 'scientific profile' (good at maths, bad at French).

In the second case (graphs 1 and 3), in contrast, the graphs show partial clouds of different shapes. For the biometric measurement, A is particular (tall and heavy) whereas for the grades it is B ('scientific profile') which is different.

More generally, we compare N_I^j in terms of their shape (we define the shape of a cloud of points by all of its interpoint distances).

Direct visual comparison of these clouds is easy, in theory, if they are each only defined by two variables. But even in this case, if there are many individuals and/or groups, a graphical tool can be useful. Procrustes analysis provides such a graphical tool. This method:

- After standardisation by the total inertia (if appropriate)
- Superimposes the configurations
- Turns them one by one, operating symmetries where appropriate, in order to bring closer (when possible) the partial points related to one individual (known as homologous points); elementary rotation which fits one cloud to another is known as *Procrustes rotation*.

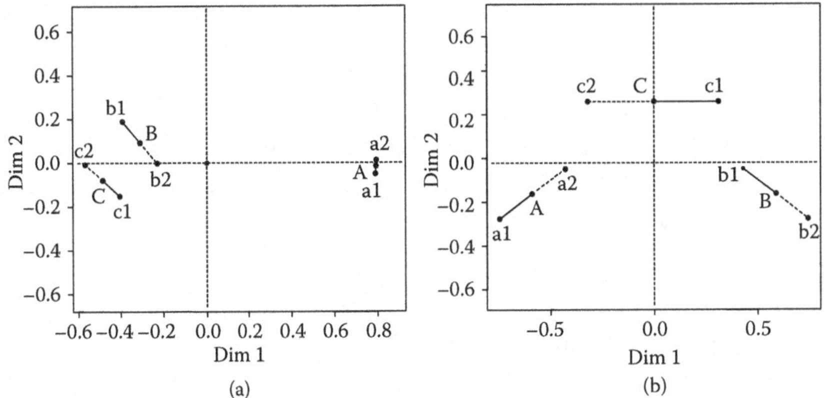

FIGURE 5.2
Procrustes analysis of the data in Figure 5.1. A: graphs 1 and 2; B: graphs 1 and 3.

We thus obtain a superimposed representation of partial clouds highlighting the similarities and differences in shape between the clouds as well as possible (that is to say in terms of a given criterion).

The results of the Procrustes analysis applied to the two above cases are given in Figure 5.2. In the first case, the similarity of the shapes of the two clouds is clear.

Chapter 9 gives more information on Procrustes analysis. In the present chapter, references to Procrustes analysis are useful for defining our objective: to construct graphs featuring partial points, or in other words, each individual as seen by each group of variables. This chapter describes MFA's solution to this problem.

5.2 Method

The individuals' space $\mathbb{R}^K$ can be seen as a direct sum of spaces $\mathbb{R}^{K_j}$:

$$\mathbb{R}^K = \bigoplus_{j=1}^{J} \mathbb{R}^{K_j}.$$

Space $\mathbb{R}^{K_j}$ used above to define N_I^j is isomorphic to subspace $\mathbb{R}^{K_j}$ of $\mathbb{R}^K$ (see Figure 5.3). In MFA, $\mathbb{R}^K$ is endowed with the diagonal metric M containing the opposite of the first eigenvalue of the separate PCAs for the groups of variables (in $\mathbb{R}^{K_j}$). This metric, being constant within a group of variables, does not question this isomorphism.

Table X_j contains the coordinates of N_I in subspace $\mathbb{R}^{K_j}$ of $\mathbb{R}^K$. As subspaces $\mathbb{R}^{K_j}$ are orthogonal by construction, cloud N_I^j is the orthogonal projection of

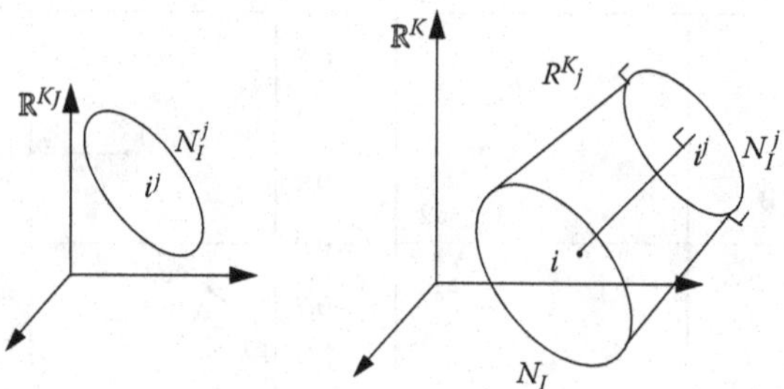

FIGURE 5.3
The partial cloud N_I^j in $\mathbb{R}^{K_j}$ and in $\mathbb{R}^K$.

N_I on $\mathbb{R}^{K_j}$. We denote $\tilde{X}_j$ the matrix X_j with added 0 to be the dimension of X, that is to say the matrix of the coordinates of N_I^j in $\mathbb{R}^K$ (variables are centred). Figure 5.4 illustrates the relative positions of N_I and N_I^j in the borderline case of two groups each containing one variable.

The complete individual i is the 'sum' of partial individuals i^j. In the graphs, it is helpful to feature i at the centre of gravity of the J points i^j. Also, in practice, clouds N_I^j are dilated with the J coefficient to generate the graphs. This is why i is known as a *mean individual.*

In this space, the representation of the individuals is obtained (in MFA) by a weighted factorial analysis of N_I. The representation of N_I^j is obtained by projecting (N_I^j) on the main axes of N_I. This procedure has several important properties for the user.

Property 1
This representation is obtained by projection, a common approach for users of factorial analysis. The geometric interpretation is clear and poses no algorithimic problems.

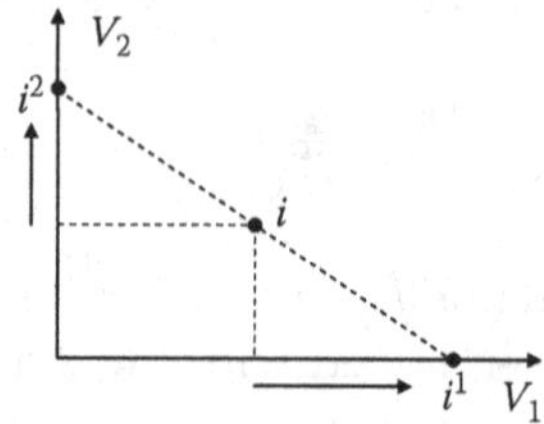

FIGURE 5.4
Complete (i) and partial (i^1 and i^2) representations of the same individual i. The first (and, respectively, second) group is reduced to variable V_1 (and V_2, respectively). From the coordinates of i, partial individuals are dilated with the J coefficient.

Property 2
This representation uses factorial axes N_I which have already been interpreted. N_I and N_I^j are analysed within the same context. This is highly important for users and in practice proves a decisive advantage (compared to a methodology implementing different methods and thus generating different factorial planes depending on the various aspects of an issue):

- Firstly in terms of time and energy spent on the interpretation (only one system of axes is interpreted)
- Then in terms of conclusions: when faced with many planes of representation from which one should we draw our conclusions? That of the weighted PCA on N_I or that of the Procrustes analyses on N_I^j?

Property 3
Using the principal axes of N_I induces relations of duality (also known as transition relations) between what is happening in $\mathbb{R}^K$ on the one hand, and in $\mathbb{R}^I$. This property is described below.

Let F_s^j be the vector of the coordinates of points i^j on the axis of rank s. By definition:

$$F_s^j = \tilde{X}_j M u_s.$$

Due to duality (in the weighted PCA of N_I):

$$u_s = \frac{1}{\sqrt{\lambda_s}} G_s.$$

By combining these two equations, we obtain:

$$F_s^j = \frac{1}{\sqrt{\lambda_s}} \tilde{X}_j M G_s.$$

The coordinates of the partial points can therefore be calculated from the variables' coordinates. This relationship, for partial individual i^j, is expressed very simply due to the unusual structure of matrix $\tilde{X}_j$ which contains only 0, except for the data related to group j. Therefore, by denoting the coordinate of i^j on u_s by either $F_s^j(i)$ or $F_s(i^j)$:

$$F_s^j(i) = F_s\left(i^j\right) = \frac{1}{\sqrt{\lambda_s}} \frac{1}{\sqrt{\lambda_1^j}} \sum_{k \in K_j} x_{ik} G_s(k).$$

Thus, for the axis of rank s, up to a coefficient (the same for all the individuals), the coordinate of i^j is the mean of the coordinates of the variables of group j weighted by the values (centred and generally reduced) of i for these variables (these weights can also be negative). In other words, partial individuals (relative to group j) are on the side of the variables (of group j) for which they have a high value, and opposite those variables (of group j) for which they have a low value. This relationship is the restriction to the variables of group j of one of the usual transition relations of the weighted PCA of N_I (see Section 1.5.4). This is the origin of the term *partial* transition relations, a property vital for interpreting the analyses.

If we express the usual transition relation (for point i and therefore all of the variables), we easily obtain

$$F_s(i) = \sum_j F_s(i^j) = \frac{1}{J} \sum_j J F_s(i^j).$$

Therefore, by dilating the partial points (with J coefficient), point i appears at the barycentre of J partial points i^j, which makes it much easier to read the graphs. It must be noted that, when calculating this barycentre, all of the J partial individuals $\{i^j; j = 1, J\}$ are of the same weight, which is logical given the desire to balance out the influence of the groups (the weight of the variables has been taken into account in calculating the coordinate of i^j).

Naturally, the symmetric transition relation, which expresses the coordinate of a variable of group j in terms of the coordinates of the partial individuals i^j, does not exist (such a relation only exists in the PCA of group j) and indeed would be undesirable as it is incompatible with a representation of the variables of different groups on one single graph.

5.3 Application to the Six Orange Juices

For the mean individuals, Figure 5.5 is exactly the same as Figure 4.4 (the percentages of inertia, which only concern these mean individuals, are therefore identical). The partial clouds are projected onto the same factorial plane. To make it easier to read, the partial individuals are connected by a line to their corresponding mean individual. The resulting representation is known as a *star graph* and for both groups is limited to segments linking corresponding partial points.

Interpretations are mostly based on partial transition relations. They can be used to compare the partial points related to a given group directly (their coordinates are calculated from the same variables).

As an example, let us compare juices 1 and 4 which, according to their mean point, can be considered equally 'hard'. It is possible to study this similarity more closely in Figure 5.5.

- From a chemical point of view, juice 4 is 'harder' than juice 1. This can also be seen in the data (see Table 1.5). Aside from a slightly lower pH for juice 1, the other variables show that juice 4 is much harder (much higher levels of citric acid, glucose and fructose; higher titre; lower level of sucrose).

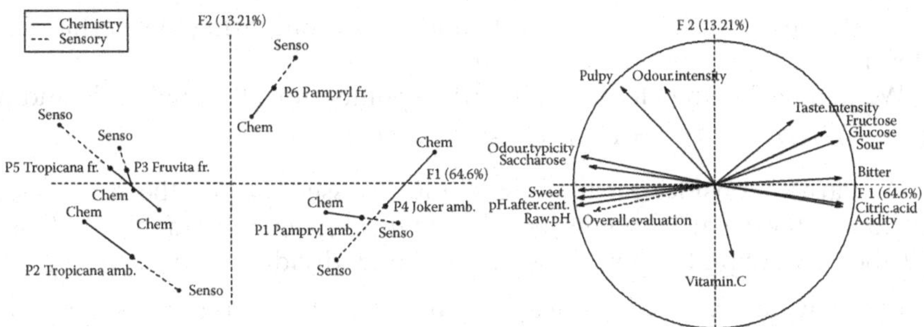

FIGURE 5.5
Orange Juice. MFA. Superimposed representation of partial clouds and the mean cloud (left). The label of a partial point is limited to the group label (Chem and Senso). Representation of variables, right (idem Figure 4.4).

- From a sensory point of view, juice 1 is 'harder' than juice 4. This observation can easily be checked in the data, with juice 1 being perceived as much more sour, more bitter and less sweet than juice 4.

The comparison of partial points related to a given group poses no problem as the partial transition relation is applied in exactly the same way as the transition relation in PCA. However, there is no property which makes it possible directly to compare partial points related to different groups inasmuch as they result from weighted means of different points (representing different variables). In such comparisons, it is therefore vital to check the data for the observations made from the factorial planes. However, when the groups' contributions to an axis are similar, as is the case for the first axis in the example, the visual impressions are generally confirmed by the data. Thus, Figure 5.5 suggests that the *chemical* hardness of juice 4 is more noteworthy than the *sensory* hardness of juice 1. This is indeed what we find in the values of the variables associated with hardness, which are more extreme for the chemical variables of juice 4 than for the sensory variables of juice 1.

We often want to compare partial points that are linked to the different groups but related to one specific individual. The graph suggests that, from the point of view of the two groups of variables, juice 1 is more homogeneous than juice 4, 'homogeneous' in the sense that the chemical and sensory hardness of juice 1 are situated closer together compared to juice 4 (high in both cases).

5.4 Interpretation Aids

It becomes difficult to analyse visually the superimposed representation of partial clouds when there are many individuals and/or groups. In such cases, it is vital to have access to indicators to select remarkable situations axis by axis. A series of indicators can be obtained by decomposing the inertia of the cloud bringing together all of the partial points.

We denote N_I^J the cloud of all partial points (for all individuals and all groups) $\{i^j; i = 1, I; j = 1, J\}$. Thus: $N_I^J = \underset{j}{\cup} N_I^j$.

We can also consider partitioning N_I^J into I classes in which the same class is attributed the J partial points associated with a given individual. We denote N_i^J the cloud of partial points associated with individual i. Thus: $N_I^J = \underset{i}{\cup} N_i^J$.

For each axis (of rank s), we apply Huygens' theorem to this second partition, decomposing total inertia (with respect to the origin O) into between-class and within-class inertia. The class mean points make up cloud N_I. The within-class inertia is that of N_i^J relative to their centre of gravity. Thus (to simplify the equations we do not mention the rank s of the axis):

$$\text{Inertia}[N_I^J/O] = \text{Inertia}[N_I/O] + \sum_i \text{Inertia}[N_i^J/i].$$

The within-class inertia of N_i^J measures the heterogeneity, along an axis u_s, of the partial points associated with individual i. By dividing it by the within-class inertia N_I^J, we obtain the (relative) contribution of individual i to this within-class inertia (for axis s). By ranking these contributions in ascending order (for each axis), we obtain:

- At the beginning, the individuals whose partial points are homogeneous for the axis in question. These individuals successfully represent the similarities between N_I^j, particularly if they are far from the origin. We select some of these to illustrate the interpretation.
- At the end, the individuals whose partial points are heterogeneous for the axis in question. They highlight differences between N_I^j which is an aspect of the variability of the data.

When there are a lot of individuals, the ranked lists (by axis) are essential in order to be able to select which individuals to examine first.

Table 5.1 brings together the contributions to the within-class inertia of the six juices. In this very small dataset, the table is not useful for interpretation (visual analysis is sufficient) and is given for information only. We can see:

- For the first axis, the heterogeneity of juices 2, 4 and 5; the homogeneity of juice 3; and the greater heterogeneity of juice 4 compared to juice 1 (described earlier)
- For the second axis, the extreme heterogeneity of juice 4

When there are more than two active groups, it can be interesting to decompose this within-individual inertia by group of variables. We thus obtain the contributions of the partial individuals to the within inertia. This

TABLE 5.1
Orange Juice. MFA. Within-Individual Inertias of the Superimposed Representation of Partial Clouds[a]

Axis 1		Axis 2	
P3 Fruvita fr.	0.47	P1 Pampryl amb.	0.38
P6 Pampryl fr.	5.36	P3 Fruvita fr.	6.15
P1 Pampryl amb.	14.43	P6 Pampryl fr.	11.93
P2 Tropicana amb.	25.19	P2 Tropicana amb.	16.22
P4 Joker amb.	26.74	P5 Tropicana fr	25.35
P5 Tropicana fr.	27.81	P4 Joker amb.	39.97
	100		100

[a] For each axis, these inertias are ranked in ascending order.

makes it possible to detect individuals which, for a group of variables, lie in an unusual position in the corresponding partial cloud compared to that which they occupy in the other partial clouds. The list of these inertias ranked in descending order is an original tool for describing the variability of a dataset which can be used to detect some anomalies and/or errors. In the case of two groups, these inertias are equal to half of the previous inertias.

If we consider all of the individuals, we can calculate the usual ratio for each axis:

$$\frac{\text{Between-class inertia}}{\text{Total inertia}} = \frac{\text{Inertia of } N_I}{\text{Inertia of } N_I^J}.$$

We thus obtain an overall indicator of the similarity between the structures of partial clouds highlighted by a given axis.

In the example (see Table 5.2), these values clearly suggest considering the first two axes in the interpretation, which was not the case for the eigenvalues. The second axis shows a similarity between the two partial clouds (high [Between inertia/Total Inertia] ratio: .6990, a far greater value than those which follow) which seems noteworthy even though it corresponds to a weak direction of inertia for N_I (low eigenvalue: .3651, barely higher than the following values).

TABLE 5.2
Orange Juice. MFA. (Between Inertia)/(Total Inertia) Ratio Associated with the Superimposed Representation

Axis 1	Axis 2	Axis 3	Axis 4	Axis 5
0.8964	0.6990	0.2046	0.3451	0.4874

Remark

The inertias of partial points cannot be added together from one axis to the other.

5.5 Distortions in Superimposed Representations

5.5.1 Example (Trapeziums Data)

We use a small example bringing together the coordinates of two trapeziums (see Table 5.3 and Figure 5.6) to illustrate our argument. Four individuals are described by two groups, each made up of two uncorrelated variables. These two variables are the same from one group to another: the only difference is the variance of the second (Y_1 and Y_2), which is accounted for in the analyses by not standardising the variables. These variables therefore play the role of principal components of separate analyses and their variances that of the corresponding eigenvalues. This is an unusual dataset as the principal components of the separate analyses are identical.

TABLE 5.3
Trapeziums Dataset

	X_1	Y_1	X_2	Y_2
a	0	0	0	0
b	10	1	10	2
c	10	3	10	6
d	0	4	0	8
Variance	25	2.5	25	10

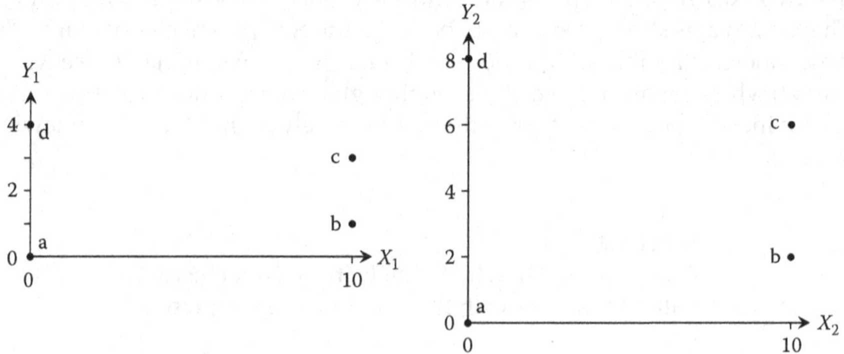

FIGURE 5.6
Trapeziums. Individuals represented for each group of variables.

TABLE 5.4
Trapeziums. Variances of the First Two Factors[a]

	F1	*F2*	*F2/F1*
Separate PCA group 1	25	2.5	0.1
Separate PCA group 2	25	10	0.4
MFA Mean Cloud	2	0.5	0.25
MFA Partial Cloud Group 1	2	0.08	0.04
MFA Partial Cloud Group 2	2	1.28	0.64

[a] In the groups analysed separately and in the representations created by the MFA. Column $F2/F1$ divides the variance of the second dimension to that of the first. This is in some ways a shape coefficient expressing the (horizontal) extension of the cloud.

Unsurprisingly (see Table 5.4 and Figure 5.7), the first axis of the MFA coincides with the two variables X_1 and X_2 and is associated with an eigenvalue of 2. The second (and final) axis coincides with the two variables Y_1 and Y_2 and is associated with an eigenvalue of 0.5; hereafter, the detailed calculus (where λ_s^j is the eigenvalue of rank s for the separate PCA of group j):

$$0.5 = \frac{2.5}{25} + \frac{10}{25} = \frac{\lambda_2^1}{\lambda_1^1} + \frac{\lambda_2^2}{\lambda_1^2}.$$

The average configuration of individuals (see Figure 5.7) is as we expect. The variance ratio between the first and second axes is, for the average configuration, the average of that for each of the separate configurations. Thus

$$\frac{0.5}{2} = \frac{1}{2}\left(\frac{2.5}{25} + \frac{10}{25}\right).$$

In the superimposed representation of the partial clouds (see Figure 5.7 and Table 5.5), we can see that the first axis is identical for the two partial clouds. The second axis shows the same opposition between individuals $\{a, b\}$ on the one hand and $\{c, d\}$, but more obviously for group 2. This corresponds to the variance of Y_2 (= 10), which is higher than that of Y_1 (= 2.5).

However, when we compare the representations of partial clouds in the MFA (see Figure 5.7) with their exact image (see Figure 5.6), we can see that the variance ratios between the two axes do not correspond. The raw data clearly show that cloud 1 has a longer shape than cloud 2. The same applies for the representation of partial clouds, but the proportions are not retained exactly. This visual impression can be quantified by comparing $F2/F1$ ratios (see Table 5.4). Compared with the raw data, the cloud of partial individuals for group 1 (and 2, respectively) is longer (and shorter, respectively). Thus, for a given axis, the representation of partial individuals in MFA respects the

TABLE 5.5
Trapeziums. MFA[a]

	F_1^1	F_2^1	F_1^2	F_2^2
a	1.4142	0.3578	1.4142	1.4311
b	−1.4142	0.1789	−1.4142	0.7155
c	−1.4142	−0.1789	−1.4142	−0.7155
d	1.4142	−0.3578	1.4142	−1.4311
Variance	2.00	0.08	2.00	1.28

[a] Coordinates of the partial individuals. F_2^1: Coordinates of the partial individuals of group 1 according to axis 2.

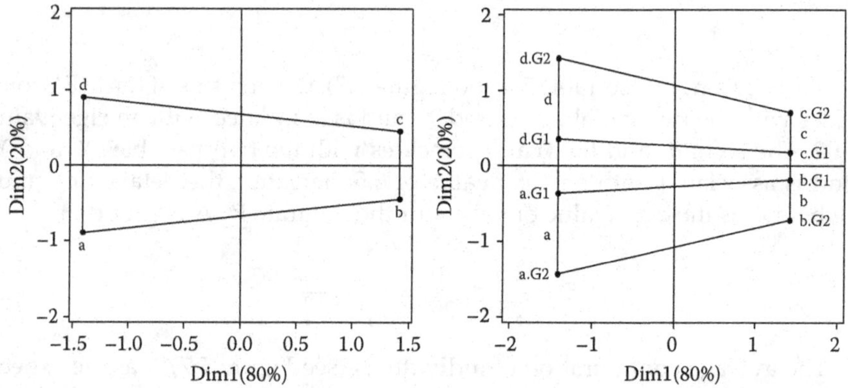

FIGURE 5.7
Mean individuals (left); mean and partial individuals (right).

relative positions of the partial points for a given group but in some ways caricatures or accentuates the differences in variability from one group to the other.

This result is general. As the dataset is small and has a particular structure, it is possible, in this specific case, to provide an exact geometric representation of it.

5.5.2 Geometric Interpretation

In $\mathbb{R}^K$ (see Figure 5.8), projected on the subspace generated by Y_1 and Y_2 (the only dimensions with nonzero inertia once the first axis has been 'removed'), cloud N_I is aligned along the axis generated by $u_2 = Y_1 + 2Y_2$, this axis being the second factorial axis. The high variance of Y_2 induces its greatest coefficient in u_2. In $\mathbb{R}^I$, vectors Y_1 and Y_2 are collinear and thus equal to the second factorial axis. Y_2, being twice as long as Y_1, has a coordinate which is also twice as

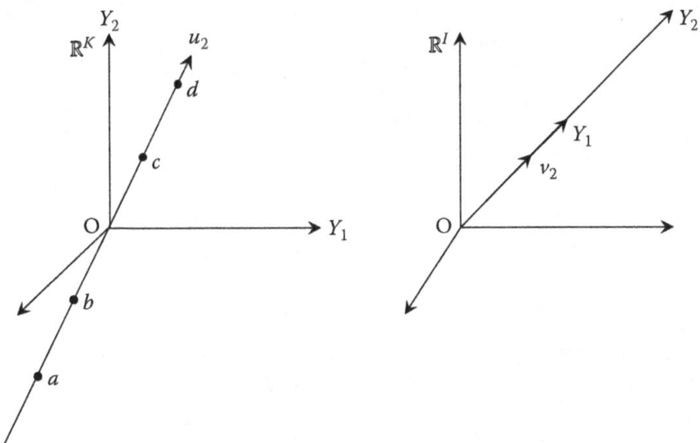

FIGURE 5.8
Duality relationship between the coordinates of u_2 (second factorial axis in $\mathbb{R}^K$) and those of Y_1 and Y_2 along v_2 (in $\mathbb{R}^I$).

big. Here we illustrate the relationship of duality in PCA (see Section 1.5.4) connecting the coordinates (in $\mathbb{R}^I$) of the variables projected on v_s (brought together in G_s) and the coordinates of u_s (in $\mathbb{R}^K$); that is to say:

$$u_s = \frac{1}{\sqrt{\lambda_s}} G_s.$$

Figure 5.9 illustrates the functioning of the superimposed representation (of the partial clouds) in MFA. The partial points are obtained by projecting the mean cloud on Y_1 (c^1, d^1) and on Y_2 $\left(c^2, d^2\right)$. In accordance with the data, the inertia of N_I^1 is lower than N_I^2 (Var $[Y_2] = 4$ Var $[Y_1]$). These clouds are then projected on u_2. As u_2 is closer to Y_2 than to Y_1, the projection reduces the inertia of N_I^1 more than that of N_I^2 (in the ratio of 1 to 4). Finally, the ratio between the variances of F_2^1 and F_2^2 (that is to say between the partial representations of groups 1 and 2 along the second axis) is 1 to 16.

The mechanism in MFA which reinforces the differences in inertia between partial clouds axis by axis is thus clarified: a partial cloud with low inertia will influence the factorial axis less and will therefore be tighter in the projection on this axis.

Remark

For point i of N_I to lie (in the graphs provided by the programs) at the centre of gravity of the corresponding partial points $\left\{i^j : j = 1, J\right\}$, the partial points need to be dilated with the J coefficient. As this coefficient is the same for all of the groups, it does not change the relative inertias of the clouds. This barycentric property is vital to the interpretation. It is therefore not possible to

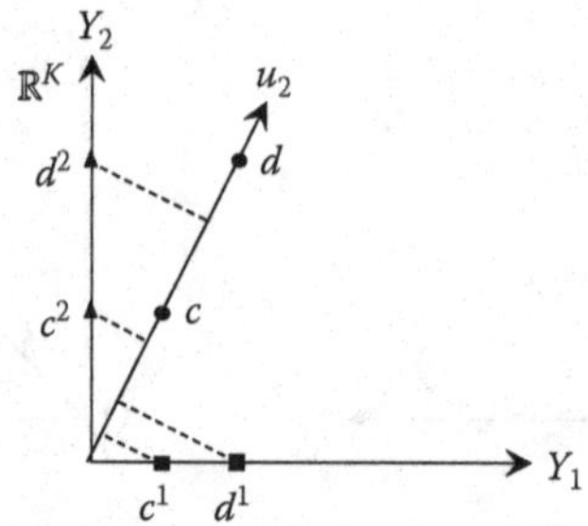

FIGURE 5.9
The partial clouds and their superimposed representation. As these clouds are symmetric with respect to the origin, we look only at individuals c and d.

reduce the distortion mentioned above (for example, by differentially dilating the partial clouds), which would mean this barycentric property would be lost.

Conclusion. For the superimposed representation we must remember that the distances between partial points are only easily interpretable for a given group of variables (expressed by the partial transition relation). From one group to the other, the distances between partial points of different groups are only meaningful if the corresponding partial clouds have comparable inertias (for the studied axes or planes).

5.5.3 Algebra Approach

Notations (Reminders and Additions)

Let x_{ik} be the general term of table X of (I, K) juxtaposing tables X_j in the rows. Let m_k be the weight attributed to variable k, M the diagonal matrix of (K, K) containing all of m_k, and M_j the diagonal matrix (K_j, K_j) containing weights m_k for the variables of group j. To simplify matters, here we attribute the same weight to all individuals.

Let $W_j = X_j M_j X_j'$ be the scalar product matrix (between individuals) associated with the jth table.

The MFA is based on a weighted PCA of table X. We denote u_s the unit vector of the sth axis of inertia of mean cloud ($u_s \in \mathbb{R}^K$), F_s the associated principal component, v_s the associated standardised principal component (F_s and $v_s \in \mathbb{R}^I$), λ_s the associated eigenvalue and S the number of nonzero eigenvalues.

Let Z be matrix (I, I) the columns of which are the standardised eigenvectors of XMX' ranked by decreasing eigenvalue; the S first columns of Z are the v_s. We obtain $Z'Z = ZZ' = $ identity.

The coordinate of partial individual i^j along the axis of rank s in the MFA is denoted $F_s(i^j) = F_s^j(i)$. F_s^j is the vector containing these I coordinates.

Furthermore, we denote $G_s(k)$ the coordinate of variable k along axis s (in $\mathbb{R}^I$) and λ_s^j the sth eigenvalue of the separate PCA of group j.

Reconstitution of N_I^j

The coordinates of the projection of N_I^j on u_s are brought together in F_s^j; F_s^j is related to v_s (see Section 5.2):

$$F_s^j = \frac{1}{\sqrt{\lambda_s}} \tilde{X}_j M G_s = \frac{1}{\sqrt{\lambda_s}} \tilde{X}_j M X' v_s = \frac{1}{\sqrt{\lambda_s}} W_j v_s.$$

Let $\{\sqrt{\lambda_s} F_s^j; s = 1, S\}$ be the set of projection coordinates of cloud N_I^j (multiplied by $\sqrt{\lambda_s}$). The PCA of the matrix (I, S) with columns $\{\sqrt{\lambda_s} F_s^j; s = 1, S\}$ is the same as that of table $W_j Z\ (I, I)$. This leads to the diagonalisation of the matrix

$$W_j Z Z' W_j = W_j W_j,$$

whose eigenvectors are the same as for W_j and the eigenvalues the squares of those of W_j.

Thus, the PCA of $\{\sqrt{\lambda_s} F_s^j; s = 1, K\}$ results in the same factors as those of X_j, the eigenvalues of this PCA being the squares of those of X_j. From this perspective, all of F_s^j enables us to reconstitute X_j. This reconstitution is not generally perfect, even if cloud N_I is perfectly represented in the MFA as, for this reconstitution:

- Factors F_s^j have already been multiplied by $\sqrt{\lambda_s}$ (distortion 1).
- The eigenvalues of the resulting reconstitution are the squares of those of the exact representation (distortion 2).

Numerical Example

In the Trapeziums dataset, the initial variables correspond to the axes of the MFA. The distortion of cloud N_I^j can therefore be illustrated by comparing the variance of an initial variable of group j (here denoted v^j) and the corresponding variance of factor F_s^j. Using the results shown above, we obtain:

$$\text{Var}\left[\frac{F_s^j}{J}\sqrt{\lambda_s}\right] = \frac{\lambda_s}{J^2}\text{Var}\left[F_s^j\right] = \left(\text{Var}\left[v^j\right]\right)^2.$$

Here, F_s^j is the vector of the coordinates provided by the programs (after the homothety of ratio J (see Section 5.2)) hence the division by J. The variance of v^j must be considered after the MFA weighting. Thus, by replacing the symbols in the second equality above by their value of group 1 and axis 2:

$$\frac{0.5}{2^2} 0.08 = \left(\frac{2.5}{25}\right)^2.$$

The squaring accurately illustrates the 'tightening' of the dimensions with low inertia. Furthermore, the effect of eigenvalue λ_s has until now not attracted our attention as it is the same for all groups. But factorial planes constructed from axes with very different inertias can show visible distortions, particularly when, initially, the corresponding dimensions of a partial cloud have equal inertias. An example of this is given in Section 9.3.2.

5.6 Superimposed Representation: Conclusion

Superimposed representation is of major interest in applications. In some cases, it makes it possible to avoid consulting separate analyses. With this in mind, we have insisted on the distortions linked to this representation. In sum, along the axis of rank s of the mean cloud, the inertia of the projection of N_I^j tends to be accentuated when the direction of projection corresponds:

- To a low direction of inertia of cloud N_I (due to distortion 1); this distortion is identical for all N_I^j
- To a high direction of inertia of N_I^j (due to distortion 2); this distortion varies according to N_I^j

In return for this distortion, the superimposed representation of N_I^j benefits from a partial transition relation expressing coordinate $F_s^j(i)$ of partial individual i^j according to the coordinates $G_s(k)$ of the variables of group j.

This relation makes it possible to take into account individuals which have data for only certain groups of variables. Such individuals cannot be introduced as active in MFA. However, we can introduce them as supplementary and represent them as partial points corresponding to the groups for which they possess data.

In the orange juice example, this would correspond to a juice for which we know the chemical variables, but which has not been tasted. Therefore, on the factorial plane of Figure 5.5, we can represent the partial point *chemical* for this juice and interpret its position compared to the other *chemical* partial points. Concretely, this will be done by completing the missing data using arbitrary values (for example, the means of the variables) and, in the results, taking into account only the partial points for which data are available. Here, MFA offers the opportunity to include data in the analysis that are usually left out.

5.7 MFA Partial Clouds in FactoMineR

We use the Orange Juice data. We continue the analysis presented in Section 4.7, explaining characteristics of partial points.

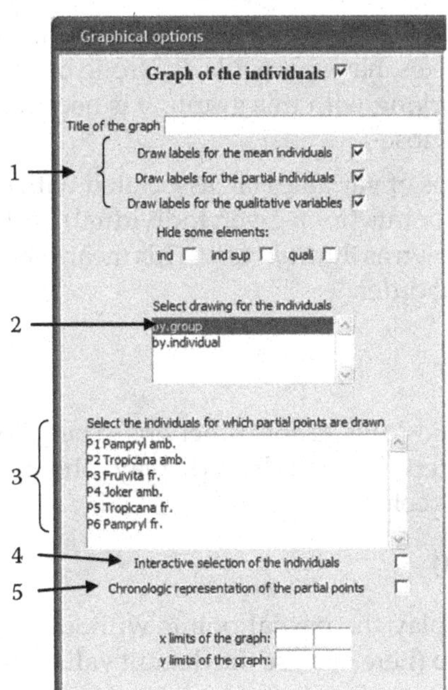

FIGURE 5.10
Individuals graph window.

The superimposed representation should contain $I(J+1)$ points and is often not easy to read as it stands. It is therefore important to be able to select the mean and/or partial individuals to be represented. We thus obtain graphs for the analysis. However, to present the results attractively, graphics software is required.

The Drop-Down Menu

Figure 5.10 shows the left part of Figure 4.10.

1. The mean and/or partial individuals can be labelled.
2. A colour can be attributed to partial individuals belonging to a given group or a given individual.
3. Individuals for which we want to represent partial points can be chosen in advance.
4. This option displays an interactive graph. At first, only the mean individuals are shown. By clicking on a mean individual, the corresponding partial individuals are shown, and these partial individuals are linked (by a segment) to their mean point (*star graph*). By clicking on a mean individual for which the partial points are displayed, the partial points become invisible. In this way, it is possible to work

step-by-step and visually to select a few notable individuals (which, for one or two axes, have particularly strong or weak within-class inertia). After working with this graph, it is necessary to close it (right click and then 'close').

5. When the groups of variables are associated with dates, it is useful to link the partial points (for a given individual) in chronological order. This is often known as the *trajectory*. This means the groups are ranked in chronological order.

Command Lines

There are two specific MFA graph functions. The `plot.MFA` function has already been presented. By default, it displays the mean individuals, each labelled in a different colour:

```
> plot.MFA(resAFM)
```

It is helpful to display the partial points without labels, but coloured according to their group (here, we use the default values of the arguments `axes` and `choix`:

```
> plot.MFA(resAFM,axes=c(1,2),choix="ind",hab="group",
+ partial="all")
```

It is possible to restrict the representation of partial individuals to a few individuals and to specify their numbers (here the first two individuals).

```
> plot.MFA(resAFM,axes=c(1,2),choix="ind",hab="group",
+ partial=c(1,2))
```

Interactive selection of the individuals for which we want to represent partial points (presented in **R Commander**) can be accessed directly using the `plot.MFApartial` function:

```
> plotMFApartial(resAFM,axes=c(1,2),hab="group")
```

This command displays the representation of mean individuals. Simply click on the points to be selected (a second click cancels the selection). When this option is controlled with **Rcmdr**, selection must always be explicitly stopped (right click or tab top left) before attempting to carry out any further operations.

Partial points are labelled by concatenating the individual's and the group's labels. Without selection, the resulting graph contains too many labels to be readable. An example is:

```
> plot(resAFM,choix="ind",partial="all",lab.par=T,hab="group")
```

One option is to display labels specific to a given graph. In order to do this, we first display a graph with no labels. We then add labels using the `text` function in the active window; an example for a graph in black and white (`habillage="none"`) is:

```
> plot(resAFM,choix="ind",partial="all",hab="none")

> text(resAFM$ind$coord.partiel[,1],resAFM$ind$coord.

+ partiel[,2],rep(c("Chim","Senso"),6),pos=3,offset=0.5)
```

The MFA output includes a lot of tables. The values of Table 5.1 are in `resAFM$ind$within.inertia`; those of Table 5.2 are in `resAFM$inertia.ratio`.

6

Factors Common to Different Groups of Variables

Historically, the simultaneous analysis of several groups of variables focused on looking for factors common to these groups. This is the aim of canonical analysis, of which there are many variations. This chapter shows how this point of view is taken into account in multiple factor analysis (MFA) and how MFA can be considered as a specific canonical analysis. We begin by describing the objectives corresponding to this new perspective, in particular the notion of *common factor*.

6.1 Objectives

A major aspect of studying a table of individuals × variables lies in identifying relationships between variables. Thus, in the simple case of one group of quantitative variables, the correlation coefficients between the variables taken two by two are examined. This analysis can also be conducted with several groups of variables, in which case we distinguish between between-group and within-group correlations. However, simultaneously taking into account several groups of variables implies that we should look at the relationship between the groups of variables themselves. The details concerning the scope of such a concept are outlined below.

This is not a new issue: in 1936, H. Hotelling published an article entitled 'Relationships Between Two Sets of Variables.' To introduce the issue, he presented the example of connecting supply and demand for raw materials in agriculture, highlighting the need to analyse all the raw materials simultaneously (for which the supplies on the one hand, and the demands, are interdependent). At the end of this section we describe the issue analysed by Hotelling in further detail. Below, however, we look at this problem in what we consider to be a more natural way, given the exploratory analysis context of this book.

Let us look back at the case of a single group of quantitative variables. Beyond the paired correlations, the relationships are studied with principal component analysis (PCA). The principal components are helpful for representing the initial variables as they maximise projected inertia. In standardised

PCA, this criterion is equivalent to the sum of the squared correlation coefficients between the principal component and the initial variables. This is why the principal components are interpreted as synthetic variables (they summarise that to which they are closely related).

In order to account for several groups of variables, we might consider implementing a PCA on the whole dataset in order to visualise both between- and within-group correlations. This was done in Chapter 4 where it was shown that, to be entirely useful, this analysis supposes that there is a balance between the groups of variables. In other words, highlighting between-group correlations from within-group correlations implicitly implies a balance between the groups of variables.

6.1.1 Measuring the Relationship between a Variable and a Group

Let us look back at the second perspective on a principal component of a standardised PCA: a component such as this is a synthetic variable closely related to the initial variables. The simplest extension of the case of several groups is to look for a variable which is closely related to the groups of variables. This requires the definition of a measurement of the relationship between a variable (here that is to say v_s, a synthetic variable of rank s) and a group of variables (already denoted K_j). Denoting such a measurement $Lg\left(v_s, K_j\right)$, the synthetic variable (of rank s) v_s must fulfill:

$$\sum_j Lg(v_s, K_j) \text{ maximum}$$

with norm constraint $\|v_s\| = 1$ and orthogonality constraint (we are, in fact, looking for a set of synthetic variables) $v_s \perp v_t$ for $t < s$.

To ensure the groups are balanced within such a criterion, it is sufficient that the measure satisfies the following relationship (as the squared correlation coefficient in standardised PCA):

$$0 \leq Lg \leq 1.$$

The value 0 indicates a total absence of relationships (in a sense which still requires clarification); the value 1 indicates a relationship of maximum intensity (in a sense which still requires clarification).

The references to the criterion of the standardised PCA (expressed as a sum of squared correlation coefficients) sheds light on what we expect from a balance between the groups of variables. In PCA, the aim is not to obtain principal components (synthetic variables) as each is linked in the same way to each variable (the possibility of obtaining such a principal component depends on the correlations between initial variables). It is merely required that, prior to the analysis, no initial variable should be privileged in constructing a principal component; this is ensured by equalising the maximum contribution to the construction of a principal component for each variable. This is indeed the case in standardised PCA, and it is precisely this property which should be transposed when simultaneously analysing several groups of variables.

6.1.2 Factors Common to Several Groups of Variables

Here we are therefore looking for the synthetic variables which are the most closely linked to the groups of variables. A synthetic variable, that is to say closely linked to each group, is referred to as a common factor (in that it is common to all groups). We can also imagine a factor common to only certain groups and even a factor specific to only one group. This would make it possible to specify the general objective of studying relationships between sets of variables: we look for factors which can be:

- Common to all groups of variables
- Common to only some groups
- Specific to one single group

Much as in PCA, once these factors have been obtained, they are studied using representations:

- Of individuals; fundamentally, a common factor is a structure on all the individuals; opposing two groups of individuals, for example.
- Of variables (correlation circle); the question is: which variables in each group correspond (that is to say, are related) to this structure on the individuals?

6.1.3 Back to the Six Orange Juices

Let us illustrate the concept of common factor using the Orange Juice data (see Section 1.10) for which we have two groups of variables: the seven sensory descriptors and the eight chemical measurements. In connecting these two types of data, we first calculate some between-group correlation coefficients for which we have an idea prior to the calculation. Thus:

- r(sour taste, pH) $= -.85$; this result is to be expected: the lower the pH is, the more acidic the solution and the more we can expect *acidic* or *sour* assessments.
- r(sweet taste, sucrose) $= .77$, this result is also expected: the higher the concentration of sucrose is, the more we expect *sweet* assessments.

It must be specified that these relationships can only be expected with all other things being equal. But, in these data, we can expect within-group relationships for two reasons.

(A) The products were chosen according to a technical variable (pasteurisation level) and their origin (Florida/other). A choice like this will lead to correlations between chemical characteristics. Thus, r(pH, sucrose) = .82: overall, the least acidic juices (chemically) have the highest sucrose content.

(B) These relationships between chemical characteristics influence several sensory variables and therefore induce relations between them. But there is more: even if we ask tasters to give a separate analysis of their perceptions, these perceptions will always be simultaneous. Considering the taste sensations only, we first perceive a taste balance, within which we try to evaluate the basic tastes. As a result, we expect relationships between sensory variables. Thus, r(sourness, sweetness) = −.90. A first idea is to connect this result to the relationship mentioned above (pH, sucrose). But we can also consider the sweet/sour taste balance (to reduce the sourness of a lemon, we often add sugar). To study the relationship between chemical and sensory data, it becomes clear in this example that we need to go beyond paired correlations (that is to say between a chemical variable and a sensory variable), and to consider the general chemical and sensory profiles (rather than their separate constituent elements). With this in mind, what does the common factor perspective have to offer?

Considering the between-group and within-group correlations amongst the four variables mentioned above (pH, sucrose, sweetness and sourness), the small number of variables (four) which, moreover, are easy to interpret (at least at first glance), and finally the small number of individuals, we might think about confronting two classes of three juices each:

1. The two Tropicana and Fruvita; these three juices have a high pH and a high sucrose content. They are perceived as sweet and slightly sour.
2. The two Pampryl and Joker; these three juices present exactly the opposite characteristics.

This opposition illustrates the concept of common factor. It is considered as a common factor in the sense that it is linked to variables from both groups (of variables). Its main advantage lies in the fact that the description of the opposition between these two classes (of juices) synthetically presents the relationships between the variables of the two groups, both between-group and within-group.

Finally, in this small example (6 individuals, 2 groups of 2 variables each), it was possible:

- To intuitively highlight this common factor without specific statistical analysis
- To validate the summary provided by the common factor by directly analysing all of the correlation coefficients

Of course this is not the case when there is a large amount of data, which is why we need a specific statistical method.

In this orange juice example, it was possible to identify a common factor thanks to the PCA results (see Section 1.10). Therefore, is PCA 'the' method for identifying common factors? No, it is not, as it is subject to an imbalance between the two groups. Moreover, in the PCA in Chapter 1, only the chemical variables are active. In this analysis, we first look for the principal dimensions of chemical variability, and only then do we connect these dimensions to the sensory variables.

6.1.4 Canonical Analysis

In order to analyse the relationships between two groups of variables, Hotelling introduces the notion of canonical variables. It simultaneously looks for:

- A linear combination (denoted u) of the variables from group 1
- A linear combination (denoted v) of the variables from group 2

such as the correlation coefficient between u and v ($r(u, v)$) is maximised. u and v are said to be *canonical variables*, $r(u, v)$ is known as the *canonical correlation coefficient* and the approach as a whole is known as *canonical analysis*.

Canonical analysis is a method which plays an important theoretical role, essentially because of the many other statistical methods which can be seen as specific cases (multiple regression, for example, if one of the groups is reduced to a single variable), but it is practically never used to process data. We believe this is because the perspective of looking for a pair of canonical variables, one in each group, does not naturally coincide with the users' questions. The orange juice example highlighted the benefits of another concept: common factor.

The idea of looking for a function linked as closely as possible to a set of groups of variables was first introduced by J. D. Carroll in 1968, as a generalisation of (Hotelling's) canonical analysis. As many generalisations of canonical analysis were put forward, it is important to specify *Carroll's generalised canonical analysis (GCA*; it is also known as *Carroll's multicanonical analysis*). Due partly to its age and partly to its focus on searching for common factors, we can consider this analysis as a benchmark method for this issue. Of course this does not mean other methods cannot be put forward, but they need to be compared to this one (as we do).

It must nonetheless be noted that in his presentation, Carroll uses that which we have referred to as *common factors* as intermediaries for calculations in order to obtain canonical variables (from a common factor, he deduces one canonical variable per group). The above presentation of Carroll's multicanonical analysis is therefore our point of view about this method.

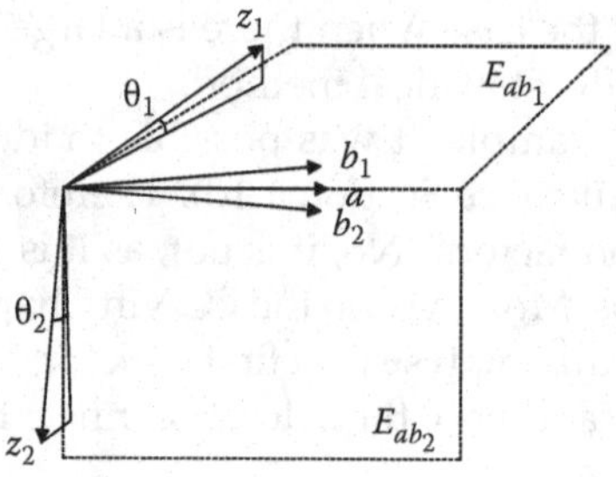

FIGURE 6.1
Geometric interpretation of the multiple correlation coefficient in an unstable situation. $\cos\theta_1$: Multiple correlation coefficient between z_1 and $\{a, b_1\}$; E_{ab_1}: Subspace generated by a and b_1.

6.2 Relationship Between a Variable and Groups of Variables

The classical measurement between a variable z and a group K_j is the multiple correlation coefficient (or its square, the determination coefficient, usually denoted R^2) denoted $r\left(z, K_j\right)$. It is the maximum correlation coefficient between z and a linear combination of variables from group K_j. Geometrically speaking, it is the cosine of the angle between z and its projection on E_j, the subspace generated by the variables of group K_j. This measurement is systematically used in multiple regression (z being the response variable; K_j containing the predictors) to assess the goodness of fit (of z by a linear combination of the variables of K_j). This measurement is entirely satisfactory when the variables K_j are not correlated with one another. In practice, we reach this situation when the variables are predictors constructed from an experimental design. Otherwise, when the variables are correlated, the subspace generated by the variables of K_j can be highly unstable (for small fluctuations of these variables) as illustrated in Figure 6.1 in an extreme situation.

In this figure, the group of variables K_j contains two variables, a and b, which are closely correlated. The variability of b is represented by means of two occurrences of b: b_1 and b_2.

In the first case (and second, respectively), group K_j, thus $\{a, b_1\}$ (and $\{a, b_2\}$, respectively), generates subspace E_{ab_1} (and E_{ab_2}, respectively). Although b_1 and b_2 are closely correlated (in other words, the variability of b represented by the deviation between b_1 and b_2 is low), the subspaces E_{ab_1} and E_{ab_2} are very different (the instability of the multiple correlation coefficient stems from its dependence on these subspaces).

The multiple correlation coefficient between z_1 (and z_2, respectively) and $\{a, b_1\}$ (and $\{a, b_2\}$, respectively) is the cosine of the angle denoted θ_1 (and θ_2, respectively) between z_1 (and z_2, respectively) and its projection on E_{ab_1} (and E_{ab_2}, respectively).

Let us examine the relationship between K_j and each of the two variables z_1 and z_2. These last two variables are nearly orthogonal to each of the variables of K_j. But the multiple correlation coefficient between z_1 and K_j is worth:

- Around 1 in the first case ($r(z_1, \{a, b_1\}) \approx 1$) as θ_1, the angle between z_1 and E_{ab_1}, is close to zero degree
- Around 1 in the second case ($r(z_1, \{a, b_2\}) \approx 0$) as θ_2, the angle between z_1 and E_{ab_2}, is close to 90°

This instability (for a slight fluctuation of b) is also observed for variable z_2. Indeed, $r(z_2, \{a, b_1\}) \approx 0$ and $r(z_2, \{a, b_2\}) \approx 1$. These two variables z_1 and z_2 are both weakly correlated to each of the variables of group K_j. A z_3 variable closely correlated to a and b (therefore correlated to b_1 and b_2) will not be affected by this instability. This is why we measure the relationship between a variable z and a group K_j, making the group of variables intervene, not only through the subspaces they generate, but also accounting for the distribution of variables within this subspace.

From this perspective, we define the measurement (of the relationship) $Lg\left(z, K_j\right)$ as the projected inertia on z of the variables of group K_j, inertia divided by its maximum value (already denoted λ_1^j, the first eigenvalue of the PCA of K_j). Thus, in the case of standardised variables (denoted v_k):

$$Lg(z, K_j) = \frac{1}{\lambda_1^j} \sum_{k \in K_j} (\text{projected inertia of } v_k \text{ on } z) = \frac{1}{\lambda_1^j} \sum_{k \in K_j} r^2(z, v_k).$$

We obtain $0 \leq Lg\left(z, K_j\right) \leq 1$.

The value 0 is reached when all the v_k variables are perfectly uncorrelated with z. This property is also true for the multiple correlation coefficient. The difference is that if the $r\left(z, v_k\right)$ are scarcely different from 0, Lg lies close to 0 whereas the multiple correlation coefficient can be high (see Figure 6.1). The value 1 is reached if z is equal to the first principal component of K_j. This first principal component has already been interpreted as the function on I which is the most closely related to the active variables of the PCA.

6.3 Searching for Common Factors

Once the measurement between a variable and a group (of variables) has been defined, the first part of Carroll's canonical analysis approach (mentioned briefly in the introduction to this chapter) is applied: look for the linear combination of variables the most closely linked to the set of groups, replacing, in Carroll's method, the squared multiple correlation coefficient by the measurement Lg. The first (common) factor, denoted v_1, is variable z which maximises

$$\sum_j Lg(z, K_j) \quad \text{with norm constraint: } \|v_1\|_D^2 = 1.$$

As $Lg(z, K_j)$ is equal to the inertia of the variables of K_j projected on z (here, inertia integrates weighting by λ_1^j), this first common factor is the first standardised principal component of the MFA. The balance of the influence of the groups, obtained thanks to the MFA's weighting, makes it possible to interpret the MFA criterion both as that of a PCA and that of a canonical analysis. This result is important in more ways than one. In particular, it shows that the two objectives, studying the variability (including several groups of variables) of individuals (the objective of PCA) on the one hand, and looking for common factors (the objective of canonical analysis), are linked to each other, or are even two aspects of the same issue. This idea has already been seen in the introduction to this chapter, in which these factors are defined as a structure on the individuals (the example given is that of the opposition between two groups each with three orange juices).

Once the first common factor has been found, we look for a second, orthogonal to the first, and so on until a sequence of orthogonal factors is found. These factors are the principal components of the MFA: that of rank s has already been denoted v_s (variance 1) or F_s (variance λ_s).

6.4 Searching for Canonical Variables

In Carroll's method, each common factor F_s is attributed a set of J canonical variables (one for each group of variables). In some ways, these variables represent the common factor F_s in each group. They are obtained by projecting F_s onto each of the subspaces generated by the variables of a group. There is therefore a homogeneity of criteria when searching for common factors and for canonical variables. In both cases, a group of variables is represented by the subspace it generates.

We adapt Carroll's approach by including the characteristics of our point of view in the canonical analysis, thus:

- Considering each group of variables while taking into account the distribution of inertia within the subspace it generates
- Highlighting the individuals' space, which is done when a common factor is defined as a structure on the individuals

This point of view suggests using the coordinates of the partial individuals (grouped together in the partial factors denoted F_s^j) as canonical variables.

Indeed, F_s^j is a structure on the individuals defined by group j (cloud N_I^j is projected) and is associated with F_s (N_I^j is projected on u_s). In order to validate this choice, it must be expressed in space $\mathbb{R}^I$, as is usually the case in canonical analysis.

Factor F_s^j was already expressed (property 3 of Section 5.2) according to the variables' coordinates in $\mathbb{R}^I$, thus (reminder: $\tilde{X}_j$ is the table X_j complemented with zeros in order to obtain the same dimensions as the complete table X):

$$F_s^j = \frac{1}{\sqrt{\lambda_s}} \tilde{X}_j M G_s.$$

In the case of standardised variables, the kth coordinate of G_s contains the correlation coefficient between the kth variable denoted v_k and F_s. In such a case:

$$F_s^j = \frac{1}{\sqrt{\lambda_s}} \frac{1}{\lambda_1^j} \sum_{k \in K_j} r\left(F_s, v_k\right) v_k.$$

Thus, F_s^j is a function on I which both accounts for the distribution of the variables of K_j (a linear combination of the variables of group K_j) and is linked to F_s (each variable v_k influences F_s^j more and more, the more closely related it is to F_s).

This interpretation of F_s^j in $\mathbb{R}^I$ validates its choice as a canonical variable in an approach inspired by that of Carroll. It makes it possible to consider MFA as a multicanonical analysis.

Remark
The idea of combining variables v_k using their correlation coefficient with variable z as coefficient is found in PLS regression (with just one component) explaining z in terms of v_k.

In practice, as has already been mentioned for the partial clouds, in order to create the graphs, the F_s^j are multiplied by J.

6.5 Interpretation Aids

The canonical analysis perspective suggests two types of supplementary indicators in order to assist in interpreting an MFA.

6.5.1 *Lg* Relationship Measurement

This is not truly a new indicator but rather an additional interpretation of the contribution of a group to the inertia of an axis. Thus, Table 4.5 now warrants the two following comments:

TABLE 6.1
Orange Juices. MFA. Canonical Correlation Coefficients

Group	F1	F2	F3	F4	F5
1: Chemical	.9466	.7556	.4407	.4631	.8030
2: Sensory	.9469	.9522	.4638	.6941	.5744

- The first factor of the MFA is closely linked to each of the groups in this way as it represents a high direction of inertia for each group (for example, $Lg(F_1, K_1) = .891$, a value close to the theoretical maximum of 1).
- The second factor of the MFA is slightly linked to the second group ($Lg(F_2, K_2) = .266$) and hardly related to the first ($Lg(F_2, K_1) = .099$).

6.5.2 Canonical Correlation Coefficients

In the original canonical analysis, that of Hotelling, a canonical correlation coefficient measures the relationship between two canonical variables of the same rank. This notion is less useful in cases with more than two groups, particularly in an approach like Carroll's.

However, it is interesting to evaluate the relationship between a common factor F_s and its representation F_s^j in group j. This relationship indicates how factor F_s can be considered as 'belonging' to group j. In order to do this, for each rank s we calculate the correlation coefficient between factor F_s and each canonical variable F_s^j.

These coefficients are said to be *canonical* in MFA. Consulted at the beginning of an interpretation, they guide the user by suggesting the type of each factor (either common to all groups, or to some of them, or specific to just one group). Applied to the Orange Juice data, these coefficients are brought together in Table 6.1.

In the absence of a validation technique, we can only empirically choose a threshold below which canonical correlation coefficients should be considered negligible. In practice, we consult these coefficients for a large number of axes, with those of the highest ranks (almost) certainly corresponding to an absence of structure; we therefore try to find discontinuity in this distribution.

Table 6.1 suggests this limit should be fixed between .8030 and .9522. This leads us to considering the first factor as common to two groups and the second factor as specific to group 2 (sensory).

This interpretation supports:

- The aforementioned relationship measurement Lg, equal to .099 (Table 4.5), very low between F_2 and group 1
- The eigenvalues of the PCA of group 1 (Table 4.5) which suggest a one-dimensional group (the first axis accounts for 77.66% of the total inertia)

– The representation of the variables on the first plane (see Figure 4.4) which suggests no chemical interpretation apart from the first axis

Methodological Remark. In practice, we begin to analyse MFA results by examining the canonical correlation coefficients. If we conclude that there are no common factors, it is logical to interrupt the analysis: in such cases, the factors of the separate PCAs will be favoured over the factors (thus specific to one group) of the MFA.

In the FactoMineR MFA output (named `res`, for example), Table 6.1 is found in `res$group$correlation`.

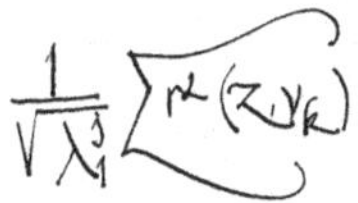

7

Comparing Groups of Variables and Indscal Model

The previous chapters present detailed tools for analysing relationships between groups of variables, both from the point of view of the variables (between-group correlations) and the individuals (comparison of partial clouds). However, when there is a large amount of data (many groups, each with a lot of variables, and a large number of individuals), more general tools are required, at least at first, in order to answer questions such as:

- Given two groups of variables, can we consider them to be related? In other words: overall, are the two associated partial clouds similar?
- As is the case for representations of individuals and variables, is it possible to generate graphs in which each group is represented by a point, with the proximity of points j and h indicating a relation/similarity between groups j and h?

7.1 Cloud N_J of Groups of Variables

The raw data associated with a group j of variables makes up table X_j. Generally, there is no correspondence between the columns of the different X_j and it is not possible to compare the X_j tables directly.

The idea of comparing two groups of variables from their partial cloud (see Chapter 5) suggests representing a group by its matrix of between-individual distances. These matrices have the same dimensions from one group to another and their entries correspond pairwise: it is therefore possible to compare them directly.

From another point of view, it is legitimate to represent a cloud of individuals by the matrix of their scalar products (between individuals; denoted XX' in Section 1.5.1) with which it is associated: indeed, diagonalising this matrix makes it possible to represent the cloud of individuals perfectly on its principal axes, with the same weight attributed to each individual (see Section 1.5.3). These two matrices are closely linked: the matrix of scalar products can be obtained by performing a dual centring of the matrix of squared distances.

The notations are as follows: $\langle i, l\rangle$ indicates the scalar product between individuals i and l; $d\,(i, l)$ the distance between i and l; $d^2\,(i, .)$ (and, respectively,

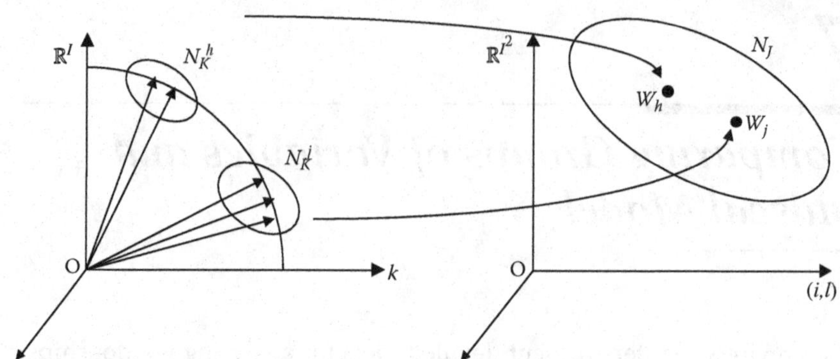

FIGURE 7.1
Clouds of variables and cloud N_J of groups of variables. Each group of variables in $\mathbb{R}^I$ is associated with a point in $\mathbb{R}^{I^2}$.

$d^2(.,l)$) the mean of the squared distances between i (and, respectively, l) and the other points. Thus

$$d^2(i,.) = \frac{1}{I}\sum_l d^2(i,l) \qquad d^2(.,l) = \frac{1}{I}\sum_i d^2(i,l).$$

$d^2(.,.)$ is the general mean of the squared between-individual distances:

$$d^2(.,.) = \frac{1}{I^2}\sum_i\sum_l d^2(i,l).$$

Torgerson's formula gives the scalar product between individuals i and l from the distances:

$$\langle i,l\rangle = \frac{1}{2}\left[d^2(i,l) - d^2(i,.) - d^2(.,l) + d^2(.,.)\right].$$

This is double centring of the matrix of squared distances by row and by column. Due to its properties, the scalar product matrix is used (rather than the distance matrix). It is traditionally denoted by the letter W; thus, for group j:

$$W_j = X_j M_j X_j'.$$

Matrix M_j represents the metric in $\mathbb{R}^{K_j}$; it is diagonal and contains the weights of the variables of group j. In MFA, this weight is generally constant within a group of variables (and equal to $1/\lambda_1^j$). Each matrix W_j contains $I \times I = I^2$ numbers; it can be attributed a point (or a vector) in a space of I^2 dimensions, denoted $\mathbb{R}^{I^2}$ and known as the *groups' space* (see Figure 7.1).

All of W_j in $\mathbb{R}^{I^2}$ make up the cloud of the groups of variables denoted N_J. In order to judge the similarity between two matrices, we examine (in $\mathbb{R}^{I^2}$)

their distance (as for the individuals in principal component analysis, PCA) and/or the angle that they form with the origin (as for the variables in PCA). It is thus necessary to endow the space $\mathbb{R}^{I^2}$ with a metric.

Each element of W_j, and therefore each dimension of $\mathbb{R}^{I^2}$, corresponds to a pair of individuals. In order to account for the individuals' weights, we associate the product $p_i p_l$ of the weights of these individuals with the dimension corresponding to the pair of individuals (i, l). The scalar product (in $\mathbb{R}^{I^2}$) between the matrices (of the scalar products between individuals, in $\mathbb{R}^K$) associated with the groups j and h is thus expressed:

$$\langle W_j, W_h \rangle_D = \sum_i \sum_l p_i p_l W_j(i,l) W_h(i,l) = \text{trace}\left(W_j D W_h D\right).$$

The D in $\langle W_j, W_h \rangle_D$ is a reminder that the weights of the individuals intervene in defining this scalar product. The justification for this metric lies in the reasoning which has already been discussed about the metric in $\mathbb{R}^I$: attributing a weight of 2 to an individual is the same as introducing this individual in the analysis twice. Therefore, attributing a weight of 2 (or 3, respectively) to individual i (and l, respectively) generates $2 \times 3 = 6$ times entry (i, l) in matrix W_j, after duplication of the individuals i and l.

Remark
It is also possible to take into account the individuals' weights, by associating matrix $W_j D$ with group j. In this case, space $\mathbb{R}^{I^2}$ is endowed with the usual metric and we obtain the same scalar product between groups j and h.

7.2 Scalar Product and Relationship Between Groups of Variables

Case of groups of variables each reduced to a single variable

$$K_1 = \{v\} \qquad K_2 = \{z\}.$$

Due to weighting in the MFA, these variables are standardised and have a weight of 1. In this case, $W_1 = vv'$ and $W_1(i,l) = v(i)\,v(l)$. The rank of this matrix is 1. An element of $\mathbb{R}^{I^2}$ is said to be *of rank 1* when it corresponds to only one element of $\mathbb{R}^I$. The scalar product between W_1 and W_2 is expressed:

$$\langle W_1, W_2 \rangle_D = \sum_i \sum_l p_i p_l v(i)\,v(l)\,z(i)\,z(l) = r^2(v,z).$$

In this simple but fundamental case, the scalar product (in $\mathbb{R}^{I^2}$) corresponds to the classic relationship measurement between two quantitative variables.

Case where only one of the two groups is multidimensional

$$K_1 = \{v\} \qquad K_2 = \{z_k : k = 1, K_2\}.$$

Variable z_k is endowed with weight m_k. When the variables are standardised, this weight is generally that of the MFA: $1/\lambda_1^2$. When the variables are only centred, we can consider reduction as a weighting (of the standardised variables) by their variance (s_k^2). Finally, in this case, their weight in MFA is s_k^2/λ_1^2.

W_2 can be written as a sum of elements of rank 1. Thus

$$W_2 = \sum_k m_k z_k z_k'.$$

This is why, by exploiting the bilinearity of the scalar product,

$$\begin{aligned}\langle W_1, W_2\rangle_D &= \left\langle vv', \sum_k m_k z_k z_k' \right\rangle_D = \sum_k m_k \left\langle vv', z_k z_k' \right\rangle_D = \sum_k m_k r^2(v, z_k)\\ &= \sum_k \text{projected intertia of } z_k \text{ on } v \text{ (in } \mathbb{R}^I) = Lg(v, K_2).\end{aligned}$$

Again here, the scalar product is interpreted as a relationship measurement (see Section 6.2).

General Case: Two multidimensional groups

$$K_1 = \{v_h : h = 1, K_1\} \qquad K_2 = \{z_k : k = 1, K_2\}.$$

Variable v_h is endowed with weight m_h and variable z_k is endowed with weight m_k.

$$W_1 = \sum_h m_h v_h v_h' \qquad W_2 = \sum_k m_k z_k z_k'$$

$$\langle W_1, W_2\rangle_D = \sum_h m_h \sum_k m_k r^2(v_h, z_k) = \sum_h m_h Lg(v_h, K_2) = \sum_k m_k Lg(z_k, K_1).$$

This scalar product in $\mathbb{R}^{I^2}$ is interpreted in space $\mathbb{R}^I$ as follows: the inertias of the variables of a group projected onto each of the variables of the other group are added together.

It is worth 0 if, and only if, each variable from one group is uncorrelated with every variable from the other group. There is no maximum value as such: this maximum depends on the dimensionality of the groups. More specifically, this quantity increases when the two groups possess a rich common structure (that is to say, when they have several common directions with high inertia in

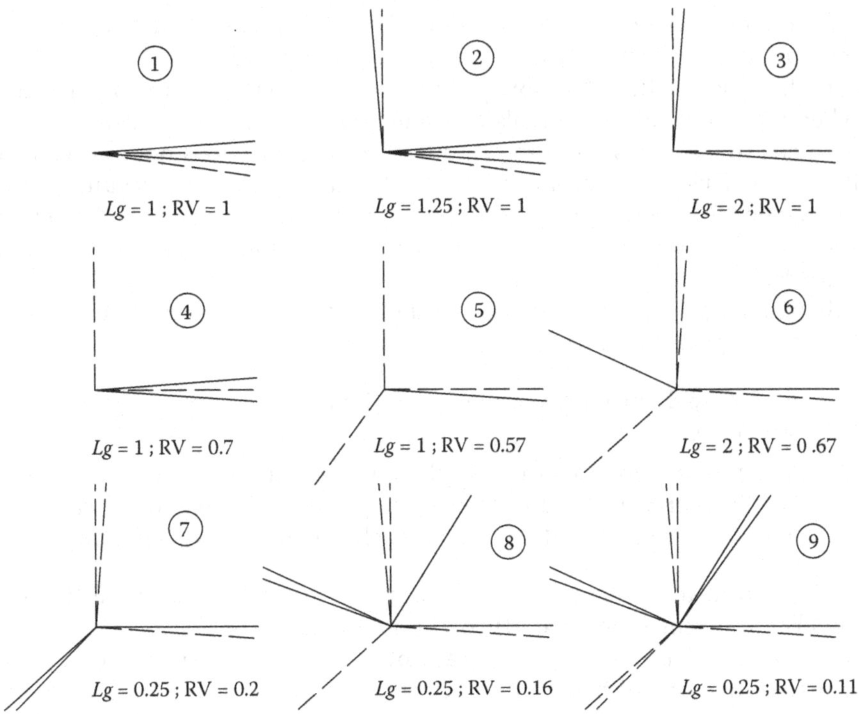

FIGURE 7.2
Lg and *RV* measurements in a few typical cases. Each segment represents a variable belonging to group 1 (continuous lines) or group 2 (broken lines). In these examples, the angles between the variables are worth 0° or 90°, which are not represented accurately on the diagrams in order to display all the variables.

each group). In this sense, we can consider this scalar product as a measurement of the relationship between two groups of variables. This aspect appears more clearly in the following section explaining the norm in $\mathbb{R}^{I^2}$.

Lg and RV
By explicitly highlighting the weighting of the MFA, the measurement Lg of the relationship between groups K_j and K_l is expressed:

$$Lg(K_j, K_l) = \left\langle \frac{W_j}{\lambda_1^j}, \frac{W_l}{\lambda_1^l} \right\rangle_D .$$

When W_1 and W_2 are standardised (in $\mathbb{R}^{I^2}$), $\langle W_1, W_2 \rangle_D$ is interpreted geometrically as a cosine and corresponds to the RV relationship measurement. Thus

$$RV(K_j, K_l) = \left\langle \frac{W_j}{\|W_j\|_D}, \frac{W_l}{\|W_l\|_D} \right\rangle_D .$$

Like measurement Lg, RV is worth 0 if, and only if, each variable of one group is uncorrelated with every variable of the other group. However, the RV coefficient is always less than or equal to 1, a value attained when the clouds of individuals associated with each of the groups are homothetic (the dimension of the clouds of individuals does not intervene in this case). This useful property makes RV the traditional measurement of the relationship between two groups of multidimensional variables. However, in cases where one of the groups is one-dimensional, Lg is more suitable.

In both measurements, we evaluate the importance of a structure common to two groups of variables:

- In RV, without paying attention to the dimension of this common structure.
- In Lg, taking into account the dimension of this common structure and its inertia relative to that of the groups. Lg is in some ways a 'number of common dimensions', each 'weighted' by its inertia.

The two measurements therefore complement each other perfectly. It is important to remember that RV or Lg is chosen according to whether we want to distinguish between the case of two one-dimensional homothetic structures and that of two multidimensional homothetic structures (in many respects, the second case is more remarkable than the first). When working with MFA, we examine both indicators.

Figure 7.2 gives the values of RV and Lg in a few typical cases explained below.

- Cases 1, 2 and 3. In these three cases, the structures generated by the two groups are homothetic: RV is always worth 1. Lg increases with the number and the inertia (relative to the first principal component of each group) of the common dimensions.
- Cases 1, 4 and 5. In these three cases, the two groups have the first principal component, and only this component, in common: Lg is always worth 1; RV decreases as the number of 'non-common' dimensions increases.
- Cases 1 and 6. Between 1 and 6, the number of common dimensions increases along with the number of non-common dimensions: Lg increases and RV decreases.
- Cases 7, 8 and 9. In these three cases, the common dimension of the two groups differs from their first principal component and, for each group, is associated with a constant inertia relative to that of the first principal component: Lg is constant, and here is worth .25; RV decreases as the number and inertia of non-common dimensions increase.

7.3 Norm in the Groups' Space

In $\mathbb{R}^{I^2}$, we associate a norm with the scalar product defined above. The square of this norm, denoted Ng, is used as an indicator of dimensionality. Thus

$$Ng\left(W_j\right) = \left\|W_j\right\|_D^2 = \left\langle W_j, W_j\right\rangle_D = \text{trace}\left(W_j D W_j D\right) = \sum_s \left(\lambda_s^j\right)^2.$$

Due to the weighting in MFA, this norm can be written:

$$Ng\left(W_j\right) = \left\|W_j\right\|_D^2 = \frac{1}{\left(\lambda_1^j\right)^2} \sum_s \left(\lambda_s^j\right)^2 = 1 + \sum_{s=2}^{s=S} \left[\frac{\lambda_s^j}{\lambda_1^j}\right]^2.$$

The groups of variables are therefore not standardised in MFA. The norm of a group increases when it has a high number of directions of inertia equal to (in practice, neighbouring) its first eigenvalue. This is why Ng is considered an indicator of dimensionality.

Dimensionality and Dimension. The dimension of a group of variables can be defined as that of the subspace generated by the variables of the group. In this subspace, the users are only interested in the dimensions associated with high inertia. We can therefore define dimensionality as the number of these dimensions of high inertia.

In the case of the six orange juices, Ng is worth 1.039 for group 1 and 1.109 for group 2. As has already been mentioned (see comment, Table 4.5), the first group has only one notable dimension and the second has two. From this perspective, indicator Ng does indeed rank the two groups according to their dimensionality. However, this indicator suggests a deviation between the two dimensionalities which is much lower than that suggested by the sequence of eigenvalues. This is because these eigenvalues are squared (in calculating Ng), making them low compared to (the first one equal to) 1 (see Table 7.1).

7.4 Representation of Cloud N_J

7.4.1 Principle

In order to represent cloud N_J of the groups of variables, the most natural way (at least for those familiar with factorial analysis) might be to project it onto its directions of maximum inertia. The Statis method, which was the first to examine cloud N_J, uses this approach. However, the axes of $\mathbb{R}^{I^2}$ found by Statis cannot be interpreted; in other words, if two groups of variables are

TABLE 7.1
Orange Juice. Eigenvalues (λ) for the Separate PCAs for the Two Groups of Variables[a]

	Chemical			Sensory		
Axis	λ	λ^2	**Total**	λ	λ^2	**Total**
1	1	1	1	1	1	1
2	.177	.031	1.031	.281	.079	1.079
3	.059	.003	1.035	.173	.030	1.109
4	.050	.002	1.037	.018	.000	1.109
5	.002	.002	1.039	.004	.000	1.109

[a] Eigenvalues are divided by the first of them, squared and cumulated.

close along one axis and far apart along another, we do not know what brings them together and what separates them. This is why, in MFA, the axes on which N_J is projected are constrained to be elements of rank 1. In this way, they correspond to a direction of $\mathbb{R}^I$ (see Section 7.2), a direction which is interpreted by examining its correlations with the initial variables.

In MFA, cloud N_J is projected on the elements of rank 1 (in $\mathbb{R}^{I^2}$) associated with the standardised principal components of the MFA (vectors of $\mathbb{R}^I$ denoted v_s). Denoting w_s the axis of rank s in $\mathbb{R}^{I^2}$, we thus obtain:

$$w_s = v_s v_s'.$$

As vectors v_s are standardised and orthogonal (in $\mathbb{R}^I$), it can easily be shown that w_s are as well (in $\mathbb{R}^{I^2}$). Thus

$$\|w_s\|_D^2 = \text{trace}\left(v_s v_s' D v_s v_s' D\right) = \text{trace}\left(v_s' D v_s v_s' D v_s\right) = 1$$

$$\langle w_s, w_t \rangle_D = \text{trace}\left(v_s v_s' D v_t v_t' D\right) = \text{trace}\left(v_t' D v_s v_s' D v_t\right) = 0.$$

The coordinate of group K_j along w_s (axis of rank s) is worth

$$\langle w_s, W_j \rangle_D = Lg\left(v_s, K_j\right) = Lg\left(F_s, K_j\right).$$

This coordinate is therefore interpreted as a relationship measurement between K_j and F_s. It is important to remember that as the axes of representation of N_J are generated by the principal components, they are interpreted as such.

Figure 7.3 features the representation of the groups in the orange juice example. In this example, there are very few groups and this representation does not add much to the interpretations that have already been made. However, this figure is sufficient to illustrate the rule of interpretation.

Graphs like this are interpreted in a similar way as correlation circles: in both cases the coordinate of a point is interpreted as a relationship measurement with a maximum value of 1. But this new graph has two specificities: the

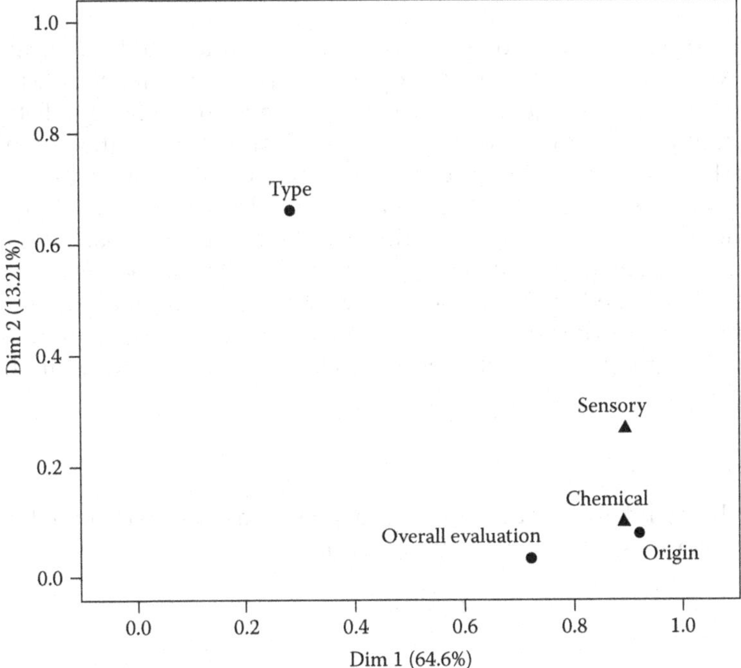

FIGURE 7.3
Orange Juice. MFA: relationship square. Representation of the groups of variables, active (triangles) and supplementary (circles).

groups of variables are unstandardised and their coordinates are always positive. They therefore appear in a square (with a side of 1 and with points [0,0] and [1,1] as vertices) known as a *relationship square* which we have already seen in the case of groups with only one variable (see Figure 3.3).

The two active groups are almost identical from the point of view of the first axis and their coordinate is high: the first axis, that is to say, the opposition between the Florida juices and the others, corresponds to a high direction of inertia in each group (which means it is correlated with several variables from each group). The two groups differ from the point of view of the second axis, as the coordinate of group 1 (chemical) is very low. The second axis (that is to say, roughly, the opposition between the refrigerated juices and the ambient juices) is invisible from a chemical perspective (at least when accounting for the available variables). On the other hand, this opposition corresponds to a sensory direction with low inertia (that is to say, it concerns few variables).

Therefore, when compared with an intuitive approach to interpretation (for a PCA user), one highly practical advantage of this representation is its relationship with the representations of individuals and variables already provided: in MFA, data are considered from different points of view, but in one single framework.

Remark

In Figure 7.3, the two qualitative variables (introduced as supplementary) were represented, as in multiple correspondence analysis (MCA; see Figure 2.3), by their squared correlation ratios with the factors. It thus appears that this first dimension more or less corresponds to the origin of the juices (Florida/other). The type of juice (ambient/refrigerated) is somewhat related to the second dimension and slightly related to the first. The overall evaluation is a supplementary group containing a single variable; its coordinates are the squared coefficient of correlation with the factors (the Florida juices are evaluated the best). These comments support the individuals' representation. Full justification for simultaneously representing quantitative and qualitative groups is described in the following chapter on qualitative and mixed data.

7.4.2 Criterion

In $\mathbb{R}^I$, the (standardised) principal component v_s maximises the quantity (with the usual orthogonality and norm constraints):

$$\sum_j Lg\left(v, K_j\right).$$

Translated in $\mathbb{R}^{I^2}$, this criterion becomes: w_s is the w which maximises the quantity

$$\sum_j \langle w, W_j \rangle_D,$$

which is the sum of the coordinates of the W_j. This quantity is maximised with, aside from the particular constraint of being of rank 1, the usual orthogonality and norm constraints:

$$\|w_s\|_D^2 = 1 \quad \text{and} \quad \langle w_s, w_t \rangle_D = 0 \quad \text{for } t < s.$$

We therefore follow the usual factorial analysis approach, that is to say, the projection of a cloud of points onto a sequence of orthogonal axes in descending order in terms of interest; however, the coordinates of the projections intervene directly in the criterion rather than through their square. This does not pose a problem, unlike in PCA for example, as these coordinates are always positive.

7.5 Interpretation Aids

Due to the criterion used, the contribution of group K_j to the axis of rank s is the coordinate of group K_j (and not its square) along the axis of rank s; we obtain a relative contribution by dividing this coordinate by the sum of the

TABLE 7.2
Orange Juice. MFA. Quality of Representation[a]

Group	Axis 1	Axis 2	Plane (1,2)	Axis 3	Axis 4	Axis 5	Ssp(1...5)
W_1	0.7657	0.0094	0.7751	0.0190	0.0038	0.0086	0.8065
W_2	0.7205	0.0640	0.7845	0.0202	0.0122	0.0022	0.8191
N_J	0.7423	0.0376	0.7799	0.0196	0.0081	0.0053	0.8130

[a] Of the groups of variables and of N_J (in $\mathbb{R}^{I^2}$). Qualities are given axis by axis, for the first plane (1,2) and for all five axes (Ssp(1...5)).

coordinates of the active groups. These contributions are interpreted in $\mathbb{R}^I$ as those of clouds N_K^j. Thus, for the relative contribution:

$$CTR(K_j, w_s) = \frac{Lg(F_s, K_j)}{\sum_j Lg(F_s, K_j)} = \frac{\text{Projected inertia of } N_K^j \text{ on } v_s}{\text{Projected inertia of } N_K \text{ on } v_s} = CTR(K_j, F_s).$$

The quality of representation of a group by an axis (or plane) can be measured as usual, using the squared cosine of the angle between the vector representing the group (W_j) and the axis (or plane). In the relationship square, it is important to check this quality because the axes are constrained to be of rank 1 (a constraint reduces the quality of the representation). Calculated for the orange juice example, these indicators are brought together in Table 7.2.

Each of the two groups possesses a quality of representation by the first plane of around .8, which can be considered entirely satisfactory. Indeed, due to the constraint of rank 1 imposed on the axes, even if all of the axes are retained (here five), the group quality of representation does not generally reach 1; the reason for this becomes clear in the section on the Indscal model.

Calculated on all of J points (last row of Table 7.2), this indicator therefore measures the quality of representation of cloud N_J by an axis or plane. This indicator differs from that calculated for clouds N_I or N_K, despite the fact that these two indicators are not independent, as one of them is based (in $\mathbb{R}^{I^2}$) on the sum of the coordinates and the other on the sum of the squared coordinates (see Tables 7.2 and 4.5). In particular, we observe the constraint effect, mentioned for each group, for cloud N_J: except in very particular cases, N_J is never perfectly represented, even when all the axes are retained.

Supplementary Elements

Supplementary groups can be introduced easily. Once the principal components (F_s) have been obtained, the coordinate of the supplementary group j along the axis of rank s, is calculated in the same way as for an active group.

It is also possible to represent isolated variables as groups with only one variable. This representation was already introduced in MCA (see Figure 2.3) and in factorial analysis of mixed data (FAMD; see Figure 3.3). It has two advantages.

For the quantitative variables, the coordinate of variable v_k along the axis of rank s (in $\mathbb{R}^{I^2}$) is worth $r^2(v_k, F_s)$. This representation of the variables is poorer than the correlation circle (the sign of the correlation coefficient is lost); however, its advantage is that it is common to both variables and groups of variables. Chapter 8 describes how to introduce the qualitative variables in this representation; in particular, geometric interpretation is presented in $\mathbb{R}^{I^2}$, for the representation of Figure 2.3 (in which the coordinate of qualitative variable q along the axis of rank s (in $\mathbb{R}^{I^2}$) is equal to the squared correlation ratio $\eta^2(q, F_s)$). This option is used in Figure 7.3, which shows that the first axis is linked to the juice's origin (Florida/other) and the second to its type (refrigerated/ambient), which we can easily find in the representation of individuals (see Figure 4.4).

7.6 The Indscal Model

The Indscal model is useful for analysing a set of distance matrices. It is expressed directly in space $\mathbb{R}^{I^2}$. In addition, the analysis of cloud N_J included in the MFA corresponds to this model. Hence the introduction of this model in this chapter.

7.6.1 Model

In the previous sections, each group j is represented by the shape of cloud N_I^j with which it is associated. Technically, to represent this shape, after having considered the distance matrices associated with N_I^j, the scalar product (between the individuals) matrix W_j is used. These two points of view, distances and scalar products, are brought together in the Indscal model, a model designed to analyse a set of distance matrices (relative to a given set of individuals).

In the original Indscal model data, J subjects were asked to evaluate the resemblances directly (or distances, in practice these distances are similarities; see end of the following section) between I stimuli. *Directly* here means that we do not ask each subject j for an evaluation of each characteristic of each stimulus, but only for an overall evaluation of the distance $d_j(i, l)$ for each pair of stimuli (i, l). The data are therefore made up of a set of J distance matrices (of $I \times I$ dimension).

The Indscal model is based on two assumptions:

1. The stimuli are differentiated according to S factors (or characteristics) for which there is a consensus among the subjects. If the stimuli are orange juices, an example of such a consensus would be: all of the subjects perceive juice 1 as very bitter, very sweet, and so on; juice 2

as not bitter at all, very sweet, and so on. We denote $z_s(i)$ the value of the sth factor for stimulus i; fundamentally, there is no ordinal relationship between the factors.

2. In their overall evaluation of stimuli, the subjects do not attribute the same importance to the different factors. For example, some judges give great importance to differences in sourness (perhaps because they are particularly sensitive to acidity) and very little importance to differences in bitterness; these judges might, for example, consider two juices with very different bitterness as similar, and consider juices which only differ according to their sourness as very different. Let q_s^j be the weight attributed by subject j to factor z_s.

With these assumptions, the distance perceived by subject j between the stimuli i and l is expressed:

$$d_j^2(i,l) = \sum_s q_s^j (z_s(i) - z_s(l))^2 + \text{residual}.$$

In this model, the factors z_s and the weights q_s^j are the parameters to be estimated. There are no constraints on the residuals, as the model is not designed within a classic inferential framework.

In other words, according to this model, there is a configuration of stimuli (defined by z_s) that each subject j 'sees' with its own diagonal metric. This point of view makes it possible to write the model for the scalar products. Thus, by denoting $w_j(i,l)$ the scalar product between two stimuli i and l from the point of view of subject j:

$$w_j(i,l) = \sum_s q_s^j z_s(i)\, z_s(l) + \text{residual},$$

which generates, for matrix W_j bringing together the scalar products for subject j:

$$W_j = \sum_s q_s^j z_s z_s' + \text{residual}.$$

In this equation, we recognise the decomposition, in $\mathbb{R}^{I^2}$, of W_j into a sum of symmetrical elements of rank 1. The representation of the groups of variables in the MFA can therefore be interpreted in terms of an Indscal model in which:

- The stimuli are the individuals; the subjects are the groups of variables.
- The factors z_s of the model are the standardised factors v_s (of the MFA).
- Weight q_s^j is the coordinate, in $\mathbb{R}^{I^2}$, of group j along the direction of rank s.

7.6.2 Estimating Parameters and Properties

The usual Indscal algorithm (for estimating the model's parameters) described in the original publication of the model uses an alternating method: z_s are fixed and q_s^j are estimated; these values q_s^j in turn serve to estimate z_s, and so on. In order to do this, the number of factors S must be fixed. These estimations are not nested (that is to say, the two factors of the estimation for $S = 2$ do not coincide with two of the factors of an estimation with $S > 2$). In practice, this problem is resolved by launching the algorithm for several values of S and then choosing a value of S which corresponds to a good balance between the goodness of fit and the number of factors.

The MFA, seen as a method for estimating the parameters of the Indscal model, is conducted in two steps.

Step 1. First, the z_s are estimated. These factors correspond as much as possible to high directions of inertia in each group. This idea (high inertia) does not appear explicitly in the Indscal model. It is essential in MFA and leads to a hierarchy of factors, from the most important (overall) to the subjects, to the least important. This hierarchy between the factors is of course necessary in practice, and users of Indscal's usual algorithms will obtain it by ranking the factors in ascending order according to the individuals' weights.

Step 2. As z_s are fixed, q_s^j are calculated. According to the interpretation of q_s^j as the relationship measurement $Lg\left(F_s, K_j\right)$, the weight that subject j attributes to dimension s increases the more closely the dimension corresponds to a strong direction of inertia of its configuration of the stimuli. This corresponds perfectly to the meaning of the weights in the Indscal model.

Compared with the usual algorithm, the estimation of the parameters of the Indscal model by the MFA presents five technical advantages:

1. The weights q_s^j are always positive. In comparison, the usual Indscal algorithm can provide negative weights interpreted as rounding errors if they are close to 0 but are otherwise not interpretable (in practice, these negative weights are replaced by zero).
2. The weights q_s^j have a maximum value of 1, a value reached when z_s corresponds to the direction of maximum inertia of subject j. The scale of variation of these weights is therefore fixed, which makes it possible to compare them amongst subjects, axes and even analyses.
3. The estimations of the parameters of the models which have different numbers of factors are nested (the two axes of the two-dimensional solution are the first two axes of the solutions with more than two dimensions).
4. Thanks to the two-stage approach (first find z_s and then q_s^j), supplementary subjects are introduced naturally.

5. This estimation is made in a framework common to several points of view (as seen in the previous chapters) which enrich one another in the interpretation.

Two other characteristics differentiate the MFA estimation and that of the Indscal algorithm.

In MFA, the F_s factors are orthogonal by construction. This constraint does not exist in the usual Indscal model. This constraint can seem useful at first as it is in factorial analysis. However, there are sometimes correlated factors (underlying the evaluation of the distances). The orange juices are a good example of this: the two experimental factors (origin and type) are not orthogonal and prove influential on the sensory plane. This is not a problem in factorial analysis as users can decide to interpret directions on the factorial plane other than the axes themselves (which can be done empirically, or more formally by optimal rotation of the axes). However, this is not possible with the Indscal model. This particularity, noticed since the origin, is especially clear with the geometric interpretation in $\mathbb{R}^{I^2}$: the plane generated by two elements of rank 1 (w_1 and w_2) does not include any other elements of rank 1 (other than w_1 and w_2). The result is that the usual Indscal algorithm retains its usefulness in the presence of nonorthogonal factors.

The other difference between the two approaches lies in the preliminary standardisation of the subjects: the Indscal algorithm standardises the subjects by fixing their total inertia at 1. The weighting in MFA is well suited to the Indscal model: particularly, multidimensional configuration must not be put at a disadvantage in constructing the first axis.

The Indscal model is Euclidean. The existence of a residual term makes it possible to imagine non-Euclidean individual data. However, the algorithms work from scalar product matrices, which means they require Euclidean individual data. Usually, individual data are not Euclidean. For example, when a subject is asked to evaluate successively the distances between pairs of stimuli, the distance matrix obtained does not generally verify triangular inequality; such a matrix is therefore not a distance matrix in the strictest sense but rather a dissimilarity matrix (symmetrical matrix of positive numbers with 0 on the diagonal). In this case, Euclidean data are obtained by a preprocessing: Torgerson's formula is applied to each dissimilarity matrix. The resulting matrix, called a pseudo-scalar product matrix, is diagonalised and only the factors related to positive eigenvalues are retained. It can be shown that this procedure, known as factorial analysis on distance tables (FADT), provides the best Euclidean approximation of a non-Euclidean distance matrix.

Specific case. There is one case in which the individual data are Euclidean: the napping®. In this data-collection procedure, subjects are asked to lay out the stimuli on a large sheet of paper (originally a tablecloth, or *nappe* in French, hence the term *napping*) so that the distances on the paper reflect the perceived distances. This data-collection method is used in both of the examples below.

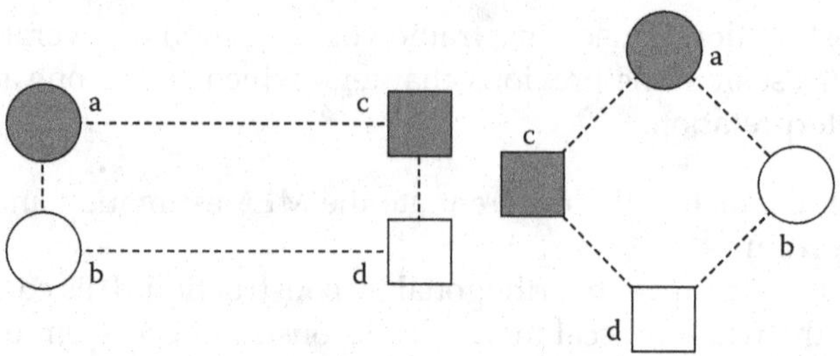

FIGURE 7.4
Cards data: four individuals (cards a, b, c and d) as seen by two groups of two variables (the data configurations by the two children).

7.6.3 Example of an Indscal model via MFA (cards)

A simple example can be used to illustrate the estimation of the parameters of the Indscal model via the MFA. There are two planar configurations of four individuals (see Figure 7.4) or, in other words, two groups of two quantitative variables each (the horizontal X and vertical Y coordinates; see Table 7.3). Here we use the MFA terminology: individuals and groups (the model stimuli and subjects, respectively).

This example is inspired by a real experiment conducted with children. Two children were presented with a set of cards (four) depicting different shapes and colours. Each child was asked to arrange the cards on the table by placing those that they considered similar close together, and placing those that they considered very different far apart. The layout of the cards was summarised using coordinates after choosing an appropriate reference.

These data were constructed using an Indscal model with two orthogonal dimensions. It is therefore unsurprising that the MFA 'finds' this model. The advantage of this example lies in the interpretation of certain results of the MFA as parameters of the Indscal model. In this MFA the variables are not standardised, so as to respect the initial distances. The estimations of the parameters are represented on a graph in Figure 7.5.

TABLE 7.3
Cards. Data[a]

	X_1	Y_1	X_2	Y_2
a	0	4	4	8
b	0	0	8	4
c	10	4	0	4
d	10	0	4	0
Variance	25	4	8	8

[a] $\{X_1, Y_1\}$: Coordinates for the left configuration (see Figure 7.4).

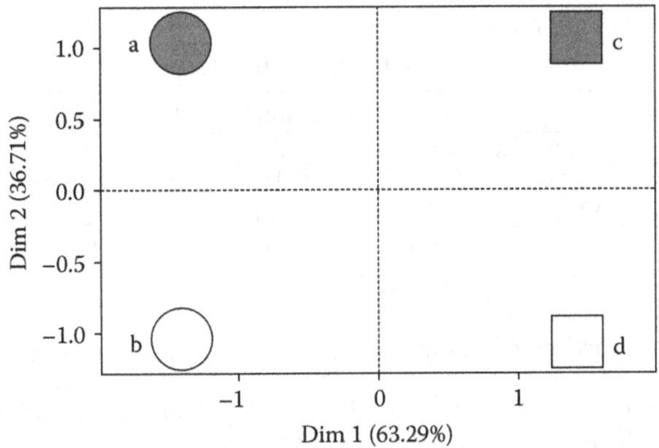

(a) Representation of individuals

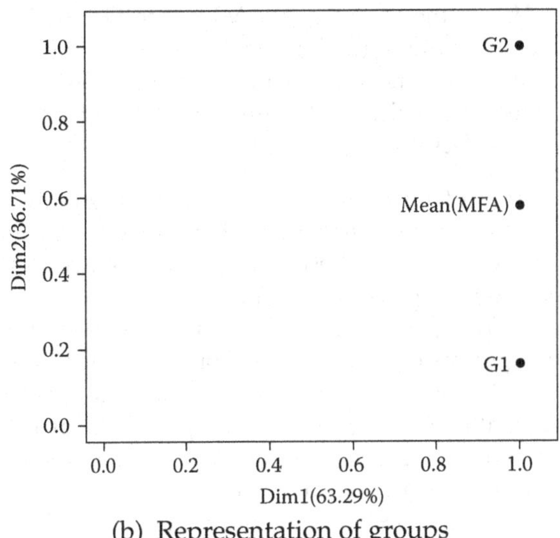

(b) Representation of groups

FIGURE 7.5
Cards. MFA. Representations corresponding to the parameters of the Indscal model. G1: group 1. Mean (MFA): set of the two groups, weighted according to the MFA.

Reminder: the whole set of coordinates of the individuals along an axis is interpreted as a factor of the Indscal model. Here these factors are interpreted simply:

- The first confronts {a,b} and {c,d} (opposition of shapes: square/circle).
- The second confronts {a,c} and {b,d} (opposition of colours: white/grey).

TABLE 7.4
Cards. Inertias in the Separate PCAs and MFA

	Total	*F1*	*F2*
PCA Group 1	29	25	4
PCA Group 2	16	8	8
MFA	3.16	2	1.16
of which group 1	1.16	1	0.16
of which group 2	2	1	1

Two representations of the individuals are possible in the Indscal model:

1. Standardised factors. This results directly from the model in the sense that the weights q_s^j are applied to the factors. This is the usual choice in the Indscal programs.
2. Unstandardised factors, with their norm measuring their relative importance. This is the choice of the MFA (see Figure 7.5), conducted independently of the Indscal model but which can be interpreted in terms of this model: the square of the norm of a factor (that is to say, the associated eigenvalue) is the sum of the weights of all of the active groups for this factor $\{q_s^j; j = 1, J\}$.

The configuration of individuals in Figure 7.5 therefore shows that the first factor is generally more important (for the subjects) than the second: its elongated shape illustrates these relative importances. Overall (here for the groups/subjects), shape is more important than colour. Finally, this configuration is a 'mean' representation of clouds associated with each group.

This relative importance can also be read on the groups' representation (see Figure 7.5). Thus, the weights for the first axis are much greater than for the second. The fact that these weights have an upper limit of 1 in MFA makes it possible to be much more precise: in this example, the first axis corresponds to the maximum direction of inertia of each group.

In practice, this group maximum is unique (the PCA rarely generates two equal first eigenvalues). From this perspective, the case of the second group is particular due to its symmetry (clearly visible in Figure 7.4): both factors are of equal importance. It should also be noted that, in the relationship square, the point of coordinates (1, 1) does not correspond to the 'mean' configuration of the individuals (see Figure 7.5) but to the standardised factors. This is why, as an interpretation aid, the point corresponding to the configuration of the individuals of Figure 7.5 is represented. This is done under the label *mean (MFA)* (for *mean configuration of MFA*). The coordinates of this point are $(1, \lambda_2/\lambda_1)$, denoting λ_1 and λ_2 the first two eigenvalues of the MFA.

Table 7.4 summarises the principal inertias in this analysis. The simplicity of these data makes it possible to find all these results easily. The eigenvalues of the separate analyses coincide with the variances of the initial variables. The groups' inertias, on the second axis of the MFA, are obtained by dividing the second eigenvalue by the first.

7.6.4 Ten Touraine White Wines

Data (Wines)

Eleven tasters (professional oenologists) were asked to evaluate 10 white wines using the napping® method. Each taster laid out the wines on a paper tablecloth with the wines that seem similar to him or her on a sensory level placed close together. Thus, each taster j is associated with a configuration of 10 wines, that is to say, a table of coordinates with 10 rows (the wines) and two columns (horizontal and vertical coordinates), except for taster 10 who used only the vertical dimension (and therefore his data have only one column). Each configuration is called a *nappe*.

These tables are horizontally brought together, all of them being subjected to an MFA in which a group corresponds to a taster (and therefore contains his or her coordinates). Here we focus on the representation of the groups (the tasters) in reference to the Indscal model. In parallel to the MFA, the parameters of this model were estimated with the usual procedure (that we refer to as Indscal, thus using the same word to designate both the model and a usual program).

Five of the wines (numbered 1 to 5) are officially labelled Touraine (Sauvignon vines). The remaining five (numbered 6 to 10) belong to the Vouvray labelling category (Chenin vines). In the latter group, two wines (7 and 10) are cask-aged (and thus have a *woody* flavour) and one (number 6) has residual sugars (7 g/L).

Figure 7.6 shows the two nappes for tasters 8 and 9. That of taster 9, for example, shows the two woody wines (7 and 10) to be relatively isolated but no separation between the Chenins and Sauvignons.

Results

The configuration of the wines provided by the MFA (see Figure 7.7), according to the first bisector, shows a separation between Chenins and Sauvignons.

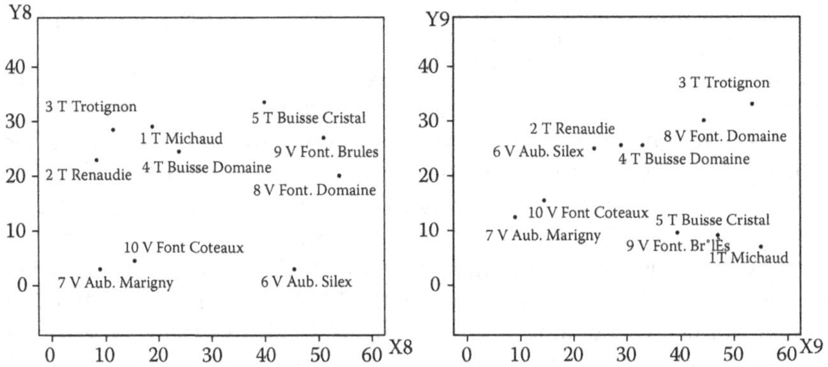

FIGURE 7.6
Wines. Nappes for tasters 8 (left) and 9 (right).

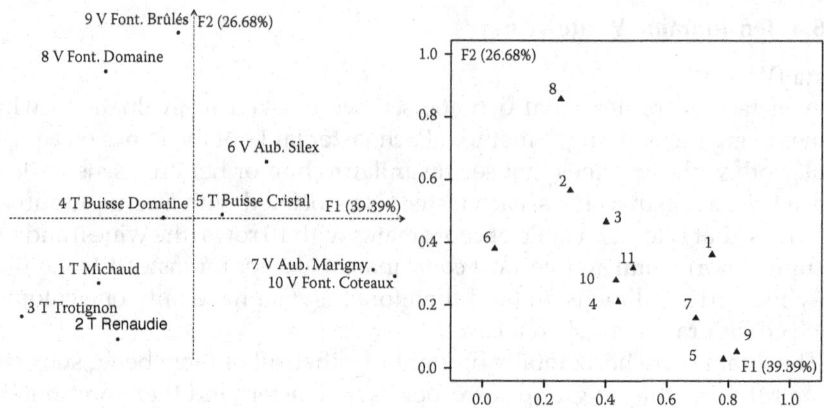

FIGURE 7.7
Wines. MFA. Representation of the wines and tasters on the first plane.

More specifically, it suggests that the Sauvignons are more homogeneous than the Chenins, and that two of them (4 and 5, from the same producer, Buisse) are less typically Sauvignon than the other three as they are closer to the centre of gravity (which is clear on the plane and remains true when we consider the overall space).

For an interpretation in terms of the Indscal model, we must look at the axes. The first axis separates the woody wines (7 and 10; these two wines make up a total contribution of 55%) from the others; it can be summarised by the term *woody*. At the other end of this axis, we find the three most typical Sauvignons: their strong fruity nose (this traditional characteristic of Sauvignons stands out from the other sensory analyses conducted on these 10 wines) can be considered, in some ways, as the opposite of *woody*.

The second axis separates wines 8 and 9 from the others (these two wines make up a total contribution of 60%). These are Chenins with no specific particularity (not cask-aged, no residual sugars). We therefore propose to summarise this axis as *Chenin typicity*.

Group representation by MFA (see Figure 7.7) is interpreted in terms of the Indscal model. For example, the high coordinate for taster 9 along axis 1 suggests that, when representing the wines, she attributed great importance to the *woody* dimension (this can be seen on her nappe, Figure 7.6, which separates 7 and 10 from the others). Another example: due to the strong coordinate along axis 2, taster 8 favoured the *Chenin typicity* dimension. Of course, this can be seen on his nappe (see Figure 7.6) on which the horizontal dimension is strongly related to the second axis on the MFA ($r(X8, F_2^{AFM}) = .93$).

When applied to the same data, the usual Indscal algorithm (for estimating the parameters of the model) generated the representations seen in Figure 7.8.

The configuration of the wines generated by Indscal is, up to a rotation, very similar to that of the MFA (*RV* between these two planar

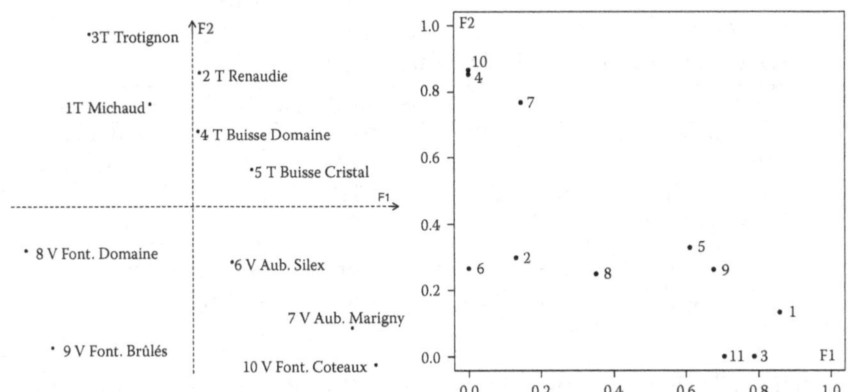

FIGURE 7.8
Wines. Indscal with two dimensions. Representation of the wines and tasters.

configurations = .95). Much like the first bisector of Figure 7.7, the second axis generated by Indscal perfectly separates the two vines. However, the first axis suggests no immediate interpretation. It clearly highlights the two woody wines, but opposes them with wines (1, 3, 8 and 9) with no clear common denominator. However, by consulting other information available for these four wines, it appears that they were often said to be *fruity*, a general and vague term (fruity odours can vary greatly) but which could be considered as opposing *woody* if *fruity* is understood as 'natural'. Finally, the second bisector almost corresponds to the first axis of the MFA ($r = .996$) but does not fit within the framework of the Indscal model.

The tasters are represented completely differently by the two methods. Indscal highlights three particular tasters (4, 7 and 10) who essentially separated the wines according to the vine, and three others (1, 3 and 11) who essentially opposed the woody and fruity wines. These two groups of tasters are mixed in the MFA representation. This example, in which the two representations of the wines differ by only one rotation and the two representations of the tasters are completely different, thus clearly illustrates why, within the context of the Indscal model, it is impossible to interpret combinations of factors, as is done in PCA, for example.

Technical Comments on the Parameters Estimated by the Indscal Program
$r(F_1^{Indscal}, F_2^{Indscal}) = -.33$: in this example, the factors on I (wines' coordinates) are loosely correlated.

Five of the tasters' coordinates are equal to 0 (tasters 4, 6 and 10 for $F1$, 3 and 11 for $F2$). The algorithm here generated negative coordinates but, as a negative weight has no meaning, they are displayed as equal to 0.

The estimations generated by Indscal, with a one- or two-dimensional model, are not nested. Nevertheless, the unique factor of the one-dimensional

TABLE 7.5
Wines. Quality of Fit of the Indscal Model[a]

	One Dimension	Two Dimensions
Indscal	.2862	.4961
MFA	.2773	.4286
Indscal without subject 10	.3118	.4860
MFA without subject 10	.3073	.4674

[a] According to the procedure (Indscal program or MFA), the number of dimensions and whether subject 10 is taken into account.

solution $F_1^{1\,\text{dim}}$ is very similar to the first factor of the first two-dimensional solution $(F_1^{2\,\text{dim}})$: $r\left(F_1^{1\,\text{dim}}, F_1^{2\,\text{dim}}\right) = .961$.

Comparing the Two Estimations: Quality of the Fit

In the usual Indscal procedure, this quality is measured by the proportion of inertia explained by the model (the sum of squares of the residuals divided by the total variability). In MFA, this quantity is interpreted as the proportion of inertia of cloud N_J expressed by an axis or plane (see Section 7.5).

Table 7.5 brings together the quality of fit of the Indscal model for different procedures. With two dimensions, this quality is slightly better (.4961 > .4286) for the Indscal estimation (see Figure 7.8) than for the MFA (see Figure 7.7). This is the expected result: the orthogonality constraint included in the MFA solution can only decrease the quality of the fit.

This is why we compare the estimation associated with only the first axis of the MFA and the Indscal estimation with just one dimension. The difference between the two procedures still favours Indscal but this time much less clearly (.2862 > .2773). Here, again, we can legitimately expect a better fit by Indscal, for which the algorithm specifically aims to optimise this fit, which is not the case in MFA.

Finally, it is important to mention that the groups are weighted differently in the two approaches: in Indscal, it is the tasters' total inertias which are standardised. Does this procedural difference explain the differences in the quality of the fit? In these data, this difference in weighting has little influence (after the MFA weighting, the tasters have more or less the same norm in $\mathbb{R}^{I^2}$) except for taster 10, whose choice to use only one dimension has already been mentioned. This is why we conduct the analyses without taster 10, which should attenuate the weighting effect. As expected (see Table 7.5), the differences still favour the Indscal procedure, but to a lesser extent (.4860 > .4674 for the two-dimensional solution and .3118 > .3073 for the one-dimensional solution).

Nature of the Factors

Although they are associated with similar qualities of fit, the estimations generated by the two procedures differ somewhat. Among other things, as

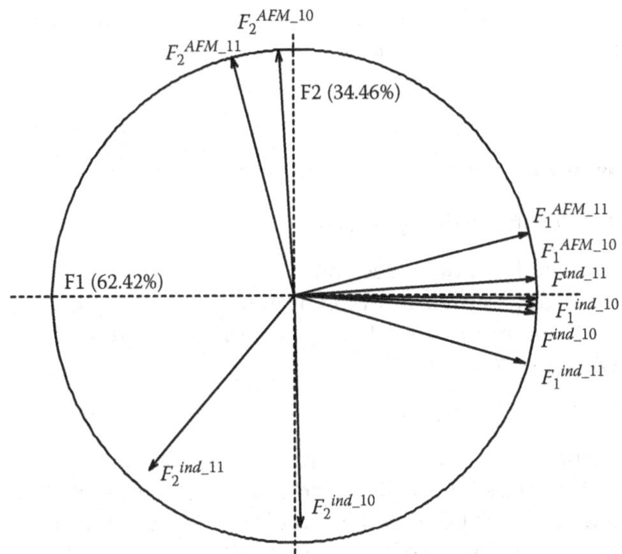

FIGURE 7.9
Wines. Standardised PCA on the factors from the different procedures. Representation of these factors on the first plane. $F_1^{ind\_10}$: first factor from Indscal with 10 subjects. F^{ind_10}: the same with a one-dimensional Indscal model.

the solutions (with different dimensions) of the Indscal procedure are not nested, it is useful to confront them with the aim of removing elements so as to improve the stability of the results. Finally, the estimation conducted without taster 10 can also provide elements of validity.

With this in mind, a standardised PCA was performed on the factors (on I) from the different procedures. It generated a representation of these factors (see Figure 7.9) with excellent representation quality (96.88%) which shows:

- (In terms of whether to take taster 10 into account) good stability of the two factors of the MFA and of the first factor of Indscal but not the second
- A notable difference between the two procedures when the 11 tasters are taken into account, but a convergence when taster 10 is eliminated
- An almost perfect nesting of the Indscal solutions when taster 10 is not included, but not otherwise

Conclusion

It is always tricky to draw general conclusions from one example. At least, we can retain that the Indscal aspect of the MFA is indeed competitive with regard to the procedure dedicated to simply estimating the parameters of the Indscal model. The main advantage of MFA here is again to provide an Indscal point of view within a general framework including many different perspectives.

Nevertheless, the usual Indscal model remains useful if we believe there may be underlying nonorthogonal factors.

7.7 MFA in FactoMineR (groups)

We use the Orange Juice data (see Table 1.5).

In **R Commander**, representations of groups in a relationship square are obtained by default for the first plane (see Figure 4.10). To obtain other planes, we use the `plot.MFA` function:

```
> plot.MFA(ResMFA,axes=c(2,3),choix="group")
```

The `MFA` function creates several tables for the groups (in $\mathbb{R}^{I^2}$): the coordinates (`res$group$coord`), the qualities of representation (`res$group$cos2`) and the total inertias (`res$group$dist2`).

Table 7.2 combines several of these results. Below, we list the R code needed to obtain it:

```
# Initialisation
> tab7_2=matrix(nrow=3,ncol=7)

# Labels of rows and columns
> row.names(tab7_2)=c("W1","W2","NJ")
> colnames(tab7_2)=c(paste("Axis",1:5),"Plane(1,2)","Ssp(1,5)")

# Quality of representation of groups
> tab7_2[1:2,1:5]=ResMFA$group$cos2[,1:5]

# Projected inertia of NJ
> tab7_2[3,1:5]=apply(ResMFA$group$coord[,1:5]^2,MARGIN=2,
  FUN=sum)

# Quality of representation of NJ
# Total inertia of the Wj are in ResMFA$group$dist2
> tab7_2[3,1:5]=tab7_2[3,1:5]/sum(ResMFA$group$dist2)

# Two column margins
> tab7_2[,6]=apply(tab7_2[1:3,1:2],MARGIN=1,FUN=sum)
> tab7_2[,7]=apply(tab7_2[1:3,1:5],MARGIN=1,FUN=sum)
> round(tab7_2,4)
```

In the relationship squares, it can sometimes be useful to add a point corresponding to all of the active tables (including the weighting of the MFA). This was done in Figure 7.5 using the *card* data. To do this, from the (active) graph displayed using the `plot.MFA` function, a point is added (`points` function): its coordinates are calculated from the eigenvalues of the MFA and its label is displayed with the `text` function.

```
# Import and verification
> Cards=read.table("Cards.csv",header=TRUE,sep=";",row.names=1)
> Cards  X1 Y1 X2 Y2
a  0  4  4  8
b  0  0  8  4
c 10  4  0  4
d 10  0  4  0

# Figure 7.4 (b)
> res=MFA(Cards,group=c(2,2),type=c("c","c"),graph=F,
> name.group=c("G1","G2"))
> plot(res,choix="group",cex=1.3)
> points(1,res$eig[2,1]/res$eig[1,1],pch=16,cex=1.3)
> text(1,res$eig[2,1]/res$eig[1,1],"mean(MFA)",offset=0.5,
+ pos=3)
```

An `indscal` function can be found in **SensoMineR**, an R package dedicated to sensory data analysis and which has also been developed by the applied mathematics department at Agrocampus. This `indscal` function is dedicated to nappes (all of the groups have two variables). Therefore, the single dimension of nappe 10 (column 19) must be duplicated. Below is the code corresponding to the analysis of the 10 Touraine white wines using the Indscal model.

```
> library(SensoMineR)
# Data importation
> napping=read.table("napping.csv",header=TRUE,sep=";",row.
  names=1)
# The unique dimension of nappe 10 is duplicated (column 19)
> res=indscal(napping[,c(1:19,19:21)])

# The Indscal function displays all the useful graphs.
# To customise the graphs in figure 7.7 (larger font)
> plot(res$points)
> text(res$points,rownames(res$points),offset=0.5,pos=3,
  cex=1.3)
> x11()
> plot(res$W)
> text(res$W,paste("",1:11,sep=""),offset=0.5,pos=3,cex=1.3)

# Table 7.5 and figure 7.8 require an Indscal solution with one
# dimension (option coord=c(1,1)). For figure 7.8, the factors
# are saved after each analysis in DonFig7_8.
# These factors are in $points (Indscal) or in $ind$coord (MFA).
> DonFig7_8=res$points
> Tab7_5=matrix(nrow=4,ncol=2)
> rownames(Tab7_5)=c("Indscal","MFA","Indscal without subject
+ 10",
```

```
# "MFA without subject 10")
> colnames(Tab7_5)=c("1 dimension","2 dimensions")

# Indscal with 1 dimension
> res1=indscal(napping[,c(1:19,19:21)],coord=c(1,1))
> DonFig7_8=cbind(DonFig7_8,res1$points)
> Tab7_5[1,]=c(res1$r2,res$r2)

# Indscal without subject 10
> res=indscal(napping[,c(1:18,20:21)])
> DonFig7_8=cbind(DonFig7_8,res$points)
> res1=indscal(napping[,c(1:18,20:21)],coord=c(1,1))
> DonFig7_8=cbind(DonFig7_8,res1$points)
> Tab7_5[3,]=c(res1$r2,res$r2)

# MFA with subject 10
> res=MFA(napping[,c(1:19,19:21)],group=rep(2,11),
+ type=rep("c",11),graph=F)
> DonFig7_8=cbind(DonFig7_8,res$ind$coord[,1:2])
> lig2=apply(res$group$coord^2,MARGIN=2,FUN=sum)/sum
+ (res$group$dist2)
> Tab7_5[2,]=c(lig2[1],sum(lig2[1:2]))

# MFA without subject 10
> res=MFA(napping[,c(1:18,20:21)],group=rep(2,10),
+ type=rep("c",10),graph=F)
> DonFig7_8=cbind(DonFig7_8,res$ind$coord[,1:2])
> lig2=apply(res$group$coord^2,MARGIN=2,FUN=sum)/sum
+ (res$group$dist2)
> Tab7_5[4,]=c(lig2[1],sum(lig2[1:2]))
> round(Tab7_5,4)

# PCA with the factors of all the analyses
> colnames(DonFig7_8)=c("F1Ind","F2Ind","FInd","F1Ind10",
+ "F2Ind10","FInd10","F1MFA","F2MFA","F1MFA10","F2MFA10")
> res=PCA(DonFig7_8)
```

$$trace(W_j D W_k D)$$

8

Qualitative and Mixed Data

Until now, multiple factor analysis (MFA) has been presented for quantitative variables. In this chapter, we extend this to qualitative variables and, more generally, to mixed data. The referent method for processing individuals × qualitative variables tables is multiple correspondence analysis (MCA). Furthermore, the issue of mixed data has already been discussed in the chapter devoted to factorial analysis of mixed data (FAMD). Here we therefore present a combination of MCA, FAMD and MFA.

Weighting the variables is key to MFA. We begin by introducing the notion of weighted variables in MCA under the name *weighted MCA.*

8.1 Weighted MCA

We use the notations for MCA as presented in Section 2.2:

- y_{ik}: General term for the complete disjunctive table (CDT); $y_{ik} \in \{0, 1\}$.
- I: The number of individuals and J the number of variables.
- K_j: The number of categories of variable j; $K = \sum_j K_j$ is the total number of categories.
- p_i: The weight of individual i. Generally, all of the individuals have the same weight, with the sum of weights fixed at 1: $p_i = 1/I$; these weights are brought together in the diagonal matrix D of dimension I: $D(i, i) = p_i$.
- p_k: The weight of the set of individuals possessing category k: $p_k = \sum_i p_i y_{ik}$. When the individuals have the same weight, p_k is the proportion of individuals possessing category k. These weights (of categories) are brought together in the diagonal matrix M of dimension K: $M(k, k) = p_k$.

In MCA, if a variable is duplicated, it is attributed a role which is twice as important. It is therefore easy to imagine an MCA in which the variables are weighted. Implicitly, in usual MCA, the weight of each variable is $1/J$. This appears most notably in the maximised criterion: the mean of squared correlation ratios.

In the weighted MCA, we denote c_j the weight attributed to variable j. To be exactly in the context of the MCA, particularly to obtain eigenvalues between 0 and 1, the sum of these weights must equal 1. We show that this condition does not apply in MFA as, for each group, the weights of the variables must ensure a maximum axial inertia of 1. But let's not get ahead of ourselves. By following the MCA presentation in chapter 2, weighted MCA is generally the same as conducting a principal component analysis (PCA) on a table X:

$$x_{ik} = \frac{y_{ik}}{p_k} - 1,$$

where individual i has a weight of p_i and category k (of variable j) of weight $m_k = p_k c_j$. Indeed, the weighted MCA only differs from usual MCA (as seen in Chapter 2), by the weighting of categories. The weighting of category k, belonging to variable j, can be seen as the product of the weighting of the category 'within' variable j (in other words p_k) by the weighting of variable j in the set of variables (that is to say, $1/J$ in usual MCA and c_j in weighted MCA), hence: $p_k c_j$.

8.1.1 Cloud of Categories in Weighted MCA

In MCA, the cloud of categories has remarkable properties. The principal characteristics of this cloud are brought together in Table 8.1, for both MCA and weighted MCA. For weighted MCA, they are obtained by replacing quantity $1/J$, in the MCA properties, by c_j.

8.1.2 Transition Relations in Weighted MCA

λ_s, F_s and G_s here designate the results (eigenvalue, coordinates of the individuals and coordinates of the categories for the axis of rank s) for MCA and weighted MCA.

In MCA and weighted MCA a category k is (up to a coefficient) at the barycentre (denoted g_k) of the individuals that possess it:

$$G_s(k) = \frac{1}{\sqrt{\lambda_s}} \frac{1}{p_k} \sum_i p_i y_{ik} F_s(i) = \frac{1}{\sqrt{\lambda_s}} F_s(g_k).$$

In MCA an individual is (up to a coefficient) at the barycentre of the categories (each attributed the same weight) that it possesses:

$$F_s(i) = \frac{1}{\sqrt{\lambda_s}} \frac{1}{J} \sum_{k \in K} y_{ik} G_s(k) = \frac{1}{\sqrt{\lambda_s}} \sum_j \frac{1}{J} \sum_{k \in K_j} y_{ik} G_s(k).$$

In weighted MCA an individual is (up to a coefficient) at the barycentre of the categories (each attributed the weight of the variable to which it belongs)

TABLE 8.1
Comparing Properties of MCA and Weighted MCA

	MCA	**Weighted MCA**
Weight of category k (of variable j)	$p_k \frac{1}{J}$	$p_k c_j$
Weight of all categories of variable j	$\frac{1}{J}$	c_j
Inertia of category k (of variable j)	$\frac{1}{J}(1 - p_k)$	$c_j(1 - p_k)$
Inertia of all categories of variable j	$\frac{1}{J}(K_j - 1)$	$c_j(K_j - 1)$
Total inertia of cloud of categories	$\sum_j \frac{1}{J} K_j - 1$	$\sum_j c_j K_j - 1$
Squared distance between categories k and h	$\sum_i p_i \left(\frac{y_{ik}}{p_k} - \frac{y_{ih}}{p_h}\right)^2$	$\sum_i p_i \left(\frac{y_{ik}}{p_k} - \frac{y_{ih}}{p_h}\right)^2$
Squared distance between individuals i and l	$\sum_j \frac{1}{J} \sum_{k \in K_j} \frac{1}{p_k}(y_{ik} - y_{lk})^2$	$\sum_j c_j \sum_{k \in K_j} \frac{1}{p_k}(y_{ik} - y_{lk})^2$

that it possesses:

$$F_s(i) = \frac{1}{\sqrt{\lambda_s}} \sum_j c_j \sum_{k \in K_j} y_{ik} G_s(k) = \frac{1}{\lambda_s} \sum_j c_j \sum_{k \in K_j} y_{ik} F_s(g_k).$$

8.2 MFA of Qualitative Variables

Within the CDT (its general term is denoted y_{ik}), the indicators are now brought together in groups which include several variables. The indicators of a given variable always belong to one group: no distinction is made between a group of indicators and a group of variables. As in all MFA, we use the letter j for the groups (group j and J the number of groups); the letter q is used for qualitative variables (variable q and Q the number of variables). In summary:

- K_j: Number of categories of group j; K: total number of categories ($K = \sum K_j$);
- Q_j: Number of variables of group j; Q: total number of variables ($Q = \sum Q_j$).

8.2.1 From the Perspective of Factorial Analysis

Principle of Weighting Groups of Variables

The issue which arises from accounting for groups in a set of variables is the same whether those variables are quantitative or qualitative: firstly the need of balancing the influence of the groups in an overall analysis, bearing in mind that a group with numerous dimensions will influence more axes than another with only few dimensions (or even one single dimension). The solution chosen by MFA (to harmonise the maximum axial inertias of the clouds associated with the different groups), previously presented within the context of quantitative variables, is not specific to these variables and applies to qualitative variables.

MFA of Qualitative Variables Is Based on a Weighted MCA

As MCA is the usual factorial analysis of a set of qualitative variables, it is natural to base the MFA of J groups of qualitative variables on a weighted MCA. In this weighted MCA, the weighting of variables, induced by the equalisation (at a value of 1) of the maximum axial inertias of the subclouds, stems from the following reasoning.

In group j prior to weighting (that is to say, in the MCA applied to group j), the distance between two individuals i and l is expressed:

$$d^2(i^j, l^j) = \frac{1}{Q_j} \sum_{k \in K_j} p_k \left(\frac{y_{ik}}{p_k} - \frac{y_{lk}}{p_k} \right)^2,$$

denoting i^j the individual i considered from the point of view of the variables of group j alone (called *partial individual* in MFA).

To make the maximum inertia of the cloud associated with group j equal to 1, we divide the weight of each category of group j by λ_1^j (first eigenvalue of the MCA of group j). After this harmonisation, the distance between i^j and l^j is expressed (see Section 2.4):

$$d^2(i^j, l^j) = \frac{1}{Q_j \lambda_1^j} \sum_{k \in K_j} p_k \left(\frac{y_{ik}}{p_k} - \frac{y_{lk}}{p_k} \right)^2 .$$

When considering all of the groups of variables, these distances generate the following distance between points i and l of N_I:

$$\begin{aligned} d^2(i,l) &= \sum_j d^2\left(i^j, l^j\right) = \sum_j \frac{1}{Q_j \lambda_1^j} \sum_{k \in K_j} p_k \left(\frac{y_{ik}}{p_k} - \frac{y_{lk}}{p_k} \right)^2 \\ &= \sum_j \sum_{k \in K_j} \frac{1}{Q_j \lambda_1^j} p_k \left(\frac{y_{ik}}{p_k} - \frac{y_{lk}}{p_k} \right)^2 . \end{aligned}$$

This relationship shows that the MFA of a set of qualitative variables organised into groups must be based on a weighted MCA in which the weighting of the variables of group j is $1/(Q_j \lambda_1^j)$.

The weight $1/(Q_j \lambda_1^j)$ of variable q for group j results from two steps:

1. Consider that variable q belongs to group j in which case the weighting is worth $1/Q_j$ in the MCA for this group.
2. Fix the maximum axial inertia of the clouds of the MCA for group j at 1 by dividing the weights of the preceding variables by λ_1^j (first eigenvalue of this MCA).

This weighted MCA provides the representations of individuals and categories in accordance with the issue: the role of the different groups is balanced. The discussion of the conditions of this balance – harmonising the maximum axial inertias, rather than the total inertias, for example – can be approached from the same perspective as for that of the quantitative variables.

Remark
When each group is made up of only one variable, $\forall j \; Q_j = 1$ and $\lambda_1^j = 1$, MFA is then equal to MCA.

In MFA, the sum of the weights of the variables is not set at 1. This is due to the fact that each group has an imposed maximum axial inertia of 1.

8.2.2 From the Perspective of Multicanonical Analysis

Lg Measurement for Qualitative Variables
For quantitative variables (see Section 6.2), it has been mentioned that MFA can be considered a multicanonical analysis as described by Carroll

(see Sections 6.2 and 6.3) on the condition that the relationship between a variable z and a group of variables $K_j = \{v_k; k = 1, K_j\}$ is measured by the Lg measurement, the projected inertia of the variables of group K_j along z. Thus

$$Lg\left(z, K_j\right) = \sum_{k \in K_j} \text{inertia of the projection of } v_k \text{ on } z.$$

When the group of variables j is made up of indicators of only one qualitative variable V, this measurement applies and it can be shown that it is worth (z is centred):

$$Lg\left(z, K_j\right) = \eta^2(z, V)$$

in which $\eta^2(z, V)$ is the squared correlation ratio between z and variable V.

If group j includes Q_j qualitative variables V_q, this measurement becomes:

$$Lg(z, K_j) = \frac{1}{\lambda_1^j} \frac{1}{Q_j} \sum_{q \in Q_j} \eta^2(z, V_q).$$

This measurement is proportional to the mean of the squared correlation ratios between variable z and the variables which make up group j. This mean is divided by its maximum value accounting for the data from group j, which is the first eigenvalue of the MCA of this group j. Indeed, in MCA, the eigenvalue of rank s is equal to the mean of the squared correlation ratios between factor F_s and the Q qualitative variables. Finally, the Lg measurement is worth 0 if the variable z has a zero correlation ratio with each variable of group j; it is worth 1 if z coincides with the principal direction of inertia of group j.

These comments are the same as those made regarding quantitative variables. They show that Lg is well suited to serve as a relationship measurement in a multicanonical analysis of qualitative variables and MFA can therefore be considered as such an analysis.

Remark

In the last relationship, $Lg(z, K_j)$ satisfies the following property: if each variable of group j is duplicated, the MCA of this group, and particularly λ_1^j, remains unchanged; the $1/Q_j$ coefficient induces the same value of $Lg(z, K_j)$ in both cases.

Looking for General Variables

In space $\mathbb{R}^I$ endowed with the metric D (containing the weights of the individuals), the general variable of rank s (denoted z_s) of this multicanonical analysis maximises:

$$\sum_j Lg(z_s, K_j) = \sum_j \frac{1}{\lambda_1^j Q_j} \sum_{q \in Q_j} \eta^2(z_s, V_q)$$

with the usual norm and orthogonality constraints:

$$\|z_s\|_D = 1 \qquad \langle z_s, z_t \rangle_D = 0 \quad \text{if } s \neq t.$$

8.2.3 Representing Partial Individuals

The notion of partial individuals, that is to say, an individual considered from the point of view of a group only, is important in MFA. We have already denoted i^j, individual i 'as seen' by group j. The geometric principle of representation of these partial individuals is identical for quantitative and qualitative variables: in space $\mathbb{R}^K$, partial individual i^j, derived from i by replacing its values for the (centred) variables by 0 for groups other than j, is projected on the axes of inertia of N_I.

The representation of the partial individuals benefits from an important property: a partial transition relation. This is inferred from the usual relationship expressing the coordinate of an individual according to that of the categories, limited to one single group of variables. These two relationships (partial and usual) are presented for qualitative variables.

In MFA on qualitative variables, the representation of N_I verifies the properties of weighted MCA (see Section 8.1.2). An individual is therefore (up to a coefficient) at the barycentre of the categories that it possesses (each attributed the weight of the variable to which it belongs). Thus

$$F_s(i) = \frac{1}{\lambda_s} \sum_j \frac{1}{\lambda_1^j Q_j} \sum_{k \in K_j} y_{ik} F_s(g_k).$$

The representation of partial individual i^j verifies the previous transition relation by restricting the sum to variables of group j (partial transition relation). Thus

$$F_s(i^j) = \frac{1}{\lambda_s} \frac{1}{\lambda_1^j Q_j} \sum_{k \in K_j} y_{ik} F_s(g_k).$$

Up to a coefficient, i^j is at the barycentre of the categories (of group j) that it possesses. This relationship can be used to compare directly the positions of points of the same cloud N_I^j (for example, i^j and l^j). However, comparisons between points from different clouds (such as i^j and l^h) are indirect: only the relative positions of these points in their respective clouds (N_I^j and N_I^h) can be compared, which only really seems useful if the factors (of MFA) being examined are common to groups j and h.

Remark

In MFA, each cloud N_I^j is dilated according to a homothety of ratio J in order to situate individual i at the centre of gravity of its partial points $\{i^j : j = 1, J\}$. The previous formula therefore becomes:

$$F_s(i^j) = \frac{1}{\lambda_s} \frac{J}{\lambda_1^j Q_j} \sum_{k \in K_j} y_{ik} F_s(g_k).$$

8.2.4 Representing Partial Categories

As categories are represented by a centre of gravity of individuals, we can imagine a concept of partial categories based on that of partial individuals. In $\mathbb{R}^K$, partial category k for group j (denoted g_k^j) is inferred from g_k by replacing its values for the variables (that is to say, the transformed and centred indicators) for groups other than j with 0. As a result, in MFA, a partial category is at the centre of gravity of the partial individuals that possess it. Thus, projecting on the axis of rank s:

$$F_s(g_k^j) = \frac{1}{p_k} \sum_i p_i y_{ik} F_s(i^j).$$

Remark

The concept of partial category is of great importance for analysing qualitative variables. In surveys, for example, there are generally a lot of individuals: they are therefore not considered individually, but instead through the subpopulations induced by their categories. This concept of *partial category* justifies the fact that, in MFA, the categories should be represented through the barycentres of their individuals rather than by indicators (or by the barycentres, dilated axis by axis by the coefficient $1/\sqrt{\lambda_s}$, which amounts to the same thing), as is usually the case in MCA.

8.2.5 Analysing in Space of Groups of Variables ($\mathbb{R}^{I^2}$)

Cloud of Groups of Variables

In $\mathbb{R}^{I^2}$, the case of qualitative variables is only slightly different from that of quantitative variables. Each group of variables K_j is represented by the matrix of the scalar products between individuals defined by group j alone. Thus, denoting M_j the diagonal matrix of dimension K_j containing the weights of the categories of group j (see also Section 7.1): $W_j = X_j M_j X_j'$. Each W_j is associated with a point in $\mathbb{R}^{I^2}$. All of these points make up the cloud of groups of variables denoted N_J.

Cloud N_J is projected on the subspace generated by the standardised principal components (of the MFA in $\mathbb{R}^I$) v_s, that is to say on $w_s = v_s v_s'$ (see Section 7.4.1). The coordinate of group j along w_s is equal to the relationship measurement $Lg\left(w_s, K_j\right)$, which is why this representation is known as a *relationship square*. $Lg\left(w_s, K_j\right)$ is also the contribution of group j to the inertia of the axis of rank s (see Section 6.5.1). Unlike for groups of quantitative variables, this Lg measurement is based on the correlation ratio (and not the correlation coefficient) for which it is a standardised mean (see Section 8.2.2).

Each qualitative variable can also be represented individually in the relationship square. The representation suggested for the MCA (see Section 2.6) is again seen here, and here has further justification (as the projection of a cloud of points).

Interpreting the Scalar Product Between Two Groups
It is important to remember that, in space $\mathbb{R}^{I^2}$, the scalar product between two groups of quantitative variables K_1 and K_2 is interpreted as a relationship measurement: RV if the groups are standardised and $Lg(K_1, K_2)$ when the maximum axial inertias of the clouds associated with the groups are equal to 1 (see Section 7.2). This property also applies to qualitative variables.

Let us remind ourselves of the two classical measurements of the relationship between two qualitative variables:

1. The χ^2 criterion, which measures the significance of the relationship via a p-value
2. The $\phi^2 = \chi^2/I$ criterion, which measures the intensity of the relationship

It can be shown that when the groups of variables K_1 and K_2 are each made up of only one qualitative variable (along with V_1 and V_2, respectively):

$$\langle W_1, W_2 \rangle_D = Lg(K_1, K_2) = \phi^2_{V_1 V_2} = \frac{1}{I}\chi^2_{V_1 V_2}.$$

In this particular case, we find the usual measurement of the intensity of the relationship between two qualitative variables. As a result, if K_1 is made up of one single variable V_1 and K_2 of Q_2 variables denoted V_q:

$$\langle W_1, W_2 \rangle_D = Lg(K_1, K_2) = \frac{1}{\lambda_1^2 Q_2} \sum_{q \in Q_2} Lg(V_1, V_q) = \frac{1}{\lambda_1^2 Q_2} \sum_{q \in Q_2} \phi^2_{V_1 V_q}.$$

The two groups are more closely related, the more closely related (in terms of ϕ^2) the variable of group 1 is to each of the variables of group 2. The same comments can be made about weighting by λ_1^2 (first eigenvalue of the MCA of group 2) as for quantitative variables. As the relationship between V_1 and K_2 increases, the closer the subspace generated by the indicators of V_1 is to the principal directions of inertia of K_2. It must be noted that, if all of the variables of group K_2 are duplicated, λ_1^2 remains unchanged. The Q_2 coefficient ensures that this relationship measurement does not vary.

Finally, if K_1 and K_2 include Q_1 and Q_2 variables, respectively (denoted V_h for group K_1 and V_q for group K_2), we obtain:

$$\langle W_1, W_2 \rangle_D = Lg(K_1, K_2) = \frac{1}{\lambda_1^1 Q_1} \frac{1}{\lambda_1^2 Q_2} \sum_{h \in Q_1} \sum_{q \in Q_2} \phi^2_{V_h V_q}.$$

The two groups are more closely related, the more closely related (in terms of ϕ^2) each variable of group 1 is to each of the variables of group 2. The same comments can be made about weighting by $1/(\lambda_1^1 \lambda_1^2 Q_1 Q_2)$ as for the previous case.

8.3 Mixed Data

The J groups of variables are now decomposed into J_1 groups of quantitative variables, J_2 groups of qualitative variables and J_3 mixed groups.

Simultaneously processing quantitative and qualitative variables by factorial analysis was addressed for FAMD (Chapter 3). But FAMD does not account for the notion of groups of variables. However, many applications require the possibility to account for variables of both kinds organised into groups. MFA offers such a possibility.

A starting point for such an analysis is to define a distance between individuals which simultaneously accounts for both types of variables. This question has already been addressed for FAMD. Concisely, it can be said that this distance must be that of the PCA if we consider only quantitative variables, and that of MCA if we consider only qualitative variables. The new element here is the structuring into groups. The balance between both types of variable must be ensured both within mixed groups (as in FAMD) and between the groups, as usual in MFA.

8.3.1 Weighting the Variables

The data table juxtaposes quantitative variables and indicators. The analysis must simultaneously take into account both types of data and function 'locally' as PCA for quantitative variables and as MCA for qualitative variables. In practice, as in FAMD, this result can be obtained directly by using the equivalence between PCA and MCA: the indicators can be processed as quantitative variables on the condition that they are appropriately weighted and coded (see Section 2.4).

The groups are balanced by the weighting of the MFA. The initial weight of variable k of group j is divided by:

- λ_1^j if group j is quantitative or mixed (λ_1^j: first eigenvalue of the PCA or the FAMD of group j)
- $\lambda_1^j Q_j$ if group j is qualitative (λ_1^j: first eigenvalue of the MCA of group j including Q_j variables (see Section 8.2.1))

The same result can be obtained by replacing the raw data with the factors of the separate factorial analysis of the groups (PCA, MCA or FAMD depending on the case). These factors are either unstandardised, or standardised and attributed an initial weight equal to their associated eigenvalue. This property suggests that the contributions (in the fullest sense of the word) to the inertias of the axes should be calculated both for the initial variables and for the factors of the separate analyses (an idea already used in Section 4.5).

8.3.2 Properties

Representing the Variables

The quantitative variables are represented with the help of their correlation coefficients with the factors. The qualitative variables essentially appear through their categories, represented by the barycentre of the individuals that possess them. In MFA, we use the exact barycentre (rather than the barycentre up to a coefficient as in MCA) due to the partial categories (for which the coefficient would not be justified). In addition, the qualitative variables themselves also appear in the relationship square.

Representing Clouds of Partial Individuals

Equalising maximum inertias of these clouds allows them to be represented simultaneously. This representation benefits from partial transition relations identical to those of the MFA applied to groups of the same type.

Up to a coefficient, partial individual i of qualitative group j lies at the barycentre of the categories of group j which it possesses; thus

$$F_s(i^j) = \frac{1}{\lambda_s} \frac{J}{\lambda_1^j Q_j} \sum_{k \in K_j} y_{ik} F_s(g_k) = \frac{1}{\sqrt{\lambda_s}} \frac{J}{\lambda_1^j Q_j} \sum_{k \in K_j} y_{ik} G_s(k).$$

Partial individual i of quantitative group j lies on the side of the variables of group j for which it has high values and opposite those of group j for which it has low values; thus

$$F_s(i^j) = \frac{1}{\sqrt{\lambda_s}} \frac{J}{\lambda_1^j} \sum_{k \in K_j} x_{ik} G_s(k).$$

The position of a partial individual of mixed group j verifies a property which combines the two preceding properties in the same way as the transition relation in FAMD (see Section 3.4). In summary, an individual lies both on the side of the quantitative variables for which it has a high value and on the side of the categories it possesses.

In the previous relationships, coefficient J ensures that individual i is at the isobarycentre of its partial points i^j, whatever their type (see also the remark at the end of Section 8.2.3). Thus

$$F_s(i) = \frac{1}{J} \sum_{j \in J} F_s(i^j).$$

Partial categories can be defined from the partial individuals. Partial category k of group j is denoted k^j; it is situated at the isobarycentre of partial individuals i^j associated with the individuals i possessing category k. This definition applies whatever the nature of group j.

Multicanonical Analysis

The quantity maximised by the factors of the MFA can be expressed in terms of a canonical analysis, with these factors as the general variables. This quantity is given hereafter, limited to three groups: K_1 (quantitative group made up of K_1 variables v_k), K_2 (qualitative group made up of Q_2 variables V_q) and K_3 (mixed group made up of K_3 quantitative variables v_k and Q_3 qualitative variables V_q). The general variable of rank s (denoted v_s) maximises

$$Lg\left(v_s, K_1\right) + Lg\left(v_s, K_2\right) + Lg\left(v_s, K_3\right)$$

$$= \frac{1}{\lambda_1^1} \sum_{k \in K_1} r^2\left(v_s, v_k\right) + \frac{1}{\lambda_1^2 Q_2} \sum_{q \in Q_2} \eta^2\left(v_s, V_q\right)$$

$$+ \frac{1}{\lambda_1^3} \left[\sum_{k \in K_3} r^2\left(v_s, v_k\right) + \sum_{q \in Q_3} \eta^2\left(v_s, V_q\right) \right]$$

with the usual norm and orthogonality constraints for v_s and denoting λ_1^j the first eigenvalue of the factorial analysis (PCA, MCA or FAMD) of group j.

As in FAMD, this quantity means that the squared correlation coefficient and the squared correlation ratio play the same role. It increases the more closely v_s is correlated with the quantitative variables and the stronger the correlation ratio with the qualitative variables. In this way, v_s is linked to K_1, K_2 and K_3. The weighting of the groups of variables limits the maximum contribution of a group to this quantity at 1. This quantity is therefore worth a maximum of 3 (general case: J), a value which is reached when v_s coincides with the first factor of the PCA of K_1 and with that of the MCA of K_2 and with that of the FAMD of K_3.

Analysis in $\mathbb{R}^{I^2}$

In this space, group j is represented by the matrix W_j of the scalar products (between individuals) that it generates. The presence of different kinds of variables poses no particular problem as the W_j matrices are homogeneous (same dimensions, same first eigenvalue). The projection of group K_j, on axis w_s generated by the general variable v_s (defined in $\mathbb{R}^I$) of the MFA, is interpreted as the relationship measurement $Lg\left(v_s, K_j\right)$ (see Section 7.4.1).

In this space, the scalar product between K_1 and K_2 is worth $Lg(K_1, K_2)$ (on the condition that the maximum axial inertia of each cloud associated with one group is 1). Depending on the nature of the variables, this scalar product derives from the correlation coefficient, from the correlation ratio or from the ϕ^2. Table 8.2 specifies $Lg(K_1, K_2)$ according to the number and the nature of the variables making up the groups in the case of homogeneous groups.

When there is both a quantitative group and a qualitative group, $Lg(K_1, K_2)$ is worth 0 if each variable from one group has a zero correlation ratio with

TABLE 8.2
Relationship Measurement $Lg\,(v_s, K_j)$[a]

Group K_1			Group K_2			
			Quantitative		**Qualitative**	
			v	$v_k, k = 1, K_2$	V	$v_k, k = 1, Q_2$
Quantitative	z		$r^2(z, v)$	$\frac{1}{\lambda_1^2} \sum_{k \in K_2} r^2(z, v_k)$	$\eta^2(z, V)$	$\frac{1}{\lambda_1^2 Q_2} \sum_{k \in Q_2} \eta^2(z, V_k)$
	$z_l, l = 1, K_1$			$\frac{1}{\lambda_1^1 \lambda_2^1} \sum_{l \in K_1} \sum_{k \in K_2} r^2(z_1, v_k)$	$\frac{1}{\lambda_1^1} \sum_{l \in K_1} \eta^2(z_1, V)$	$\frac{1}{\lambda_1^1 \lambda_1^2} \sum_{l \in K_1} \sum_{k \in Q_2} \eta^2(z_1, V_k)$
Qualitative	Z				ϕ^2_{ZV}	$\frac{1}{\lambda_1^2 Q_2} \sum_{k \in Q_2} \phi^2_{ZV_k}$
	$Z_l, l = 1, Q_1$					$\frac{1}{\lambda_1^1 Q_1} \frac{1}{\lambda_1^2 Q_2} \sum_{l \in Q_1} \sum_{k \in Q_2} \phi^2_{Z_1 V_k}$

[a] Between two groups of variables according to their type (quantitative/qualitative) and the number of variables (one/several). $r^2\,(z, v)$: Squared correlation coefficient between the quantitative variables z and v; $\eta^2(z, V)$: Squared correlation ratio between quantitative variable z and qualitative variable V; ϕ^2_{ZV}: $\phi^2 = \chi^2/I$ statistic between the qualitative variables Z and V.

each variable from the other group. There is no upper limit; it increases when both groups have many common directions of high inertia. This last property is true for all types of groups.

8.4 Application (Biometry2)

The properties of MFA as performed on mixed data are illustrated below using a small example of data chosen specifically to highlight these properties.

The biometry2 dataset is constructed as follows (see Table 8.3):

- Six individuals (A,…,F) are described by three quantitative variables (length, weight, width), two of which are correlated with each other ($r(length, width) = -.71$) and only slightly correlated to the third ($r(length, weight) = -.39\,; r(width, weight) = -.13$).
- These three variables are also coded into three qualitative variables by subdividing their variation range into two or three classes.

The specific aim of this application is to compare two images of one single dataset: that provided by the standardised data and that provided by coding into classes. The 'canonical analysis perspective' of MFA, by which the method highlights the factors common to the variables groups and those which are specific to only one of them, does indeed correspond to this comparison. The point of view 'factorial analysis of the separate factors' makes it possible to compare the two usual methodologies: standardised PCA on raw data and MCA on data coded into classes. The more general aim of this application is to show how MFA simultaneously accounts for both types of variable.

It must be noted that, in this case, there are very few individuals, particularly for an MCA. An analysis such as this would probably be of no great use with real data, but these data were constructed according to a simple and clear structure which should be highlighted with the two codings.

8.4.1 Separate Analyses

Inertias (Table 8.4)

In MCA the projected inertias are always less than 1; in PCA, the first eigenvalue is always greater than 1. Weighting the variables is vital to analyse these two types of data simultaneously.

The decrease of eigenvalues is less pronounced for the qualitative group (MCA). This is a classic observation. The first (qualitative) group is three-dimensional with the two first eigenvalues quite close together. The second (quantitative) group is two-dimensional, which fits the way it was constructed.

TABLE 8.3

Biometry2. Six Individuals (A,. . . , F) as Described by Three Quantitative Variables[a]

	Raw Data			Standardised			Coded into Classes		
	Length	Weight	Width	Length	Weight	Width	Length	Weight	Width
A	1	1	6	−1.464	0.447	1.464	1	2	3
B	2	0	5	−0.878	−0.894	0.878	1	1	3
C	3	2	3	−0.293	1.789	−0.293	2	2	2
D	5	0	4	0.878	−0.894	0.293	3	1	2
E	4	1	1	0.293	0.447	−1.464	2	2	1
F	6	0	2	1.464	−0.894	−0.878	3	1	1
Mean	3.5	0.667	3.500	0	0	0			
S.D.	1.708	0.745	1.708	1	1	1			

[a] These data are subjected to MFA through two transformations: standardising and coding into (2 or 3) classes.

TABLE 8.4
Biometry2. Eigenvalues of the Separate Analyses

	Eigenvalues				Percentages of Inertia			
Group	1	2	3	4	1	2	3	4
1 Qualitative (MCA)	0.667	0.605	0.333	0.061	40.0	36.3	20.0	3.7
2 Quantitative (PCA)	1.765	1.110	0.125		58.8	37.0	4.2	

TABLE 8.5
Biometry2. Correlations Between Factors of Separate Analyses[a]

		Group 1 (MCA)		
		F1	*F2*	*F3*
Group 2 (PCA)	*F*1	−0.82	−0.37	−0.41
	*F*2	0.43	−0.87	−0.04
	*F*3	0.33	0.27	−0.68

[a] Example: .43 is the correlation coefficient between the second factor of the PCA of group 2 and the first factor of the MCA of group 1.

Correlations Between the Factors of the Separate Analyses (Table 8.5)
Limiting ourselves to the first two, the factors of the same rank are quite closely correlated (.82 and .87): as expected, the results of the two analyses are linked. That said, higher coefficients could have been expected. In fact, these coefficients give a pessimistic view of the relationships between the results, as a similarity between two factorial planes can be masked by a rotation. Such a rotation is highly plausible in this case, because the first two eigenvalues of the MCA are close together: it is the first plane of the MCA which is stable rather than the axes themselves.

8.4.2 Inertias in the Overall Analysis

The sequence of eigenvalues (see Table 8.6) suggests interpreting three axes. The first eigenvalue of 1.920 is close to its maximum (the number of groups): the first axis of the MFA corresponds to a high direction of inertia in each group. The two groups contribute equally to this first axis, as expected

TABLE 8.6
Biometry2. MFA. Decompositions of Inertia by Axis and by Group

	Total Inertia	*F1*	*F2*	*F3*	*F4*	*F5*
Total Inertia	4.200	1.920	1.530	0.610	0.120	0.020
Group 1	2.500	0.947	0.909	0.536	0.098	0.010
Group 2	1.699	0.972	0.621	0.072	0.020	0.015

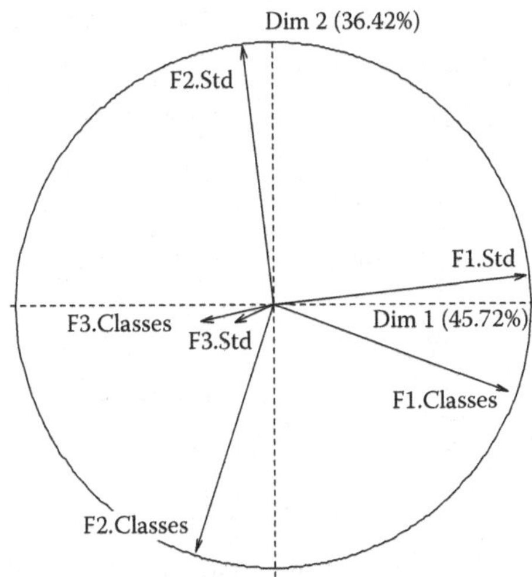

FIGURE 8.1
Biometry2. MFA. First plane. Factors of the separate analyses. Std: Standardised.

due to the weighting of the MFA (this property is almost always observed in practice).

The second axis corresponds to a direction of high inertia for group 1 and lower for group 2. This was also expected due to the first two eigenvalues of the separate analyses. The third axis is specific to group 1. This is coherent with the data construction method used.

We here illustrate the major difficulty inherent to the simultaneous factorial analysis of quantitative and qualitative variables: weighting of the MFA standardises the maximum axial inertia of each group. However, due to the decrease of the inertias which is generally lower in MCA than in PCA, the qualitative groups should be more influential, in terms of contribution to inertia, from the second axis on. This difficulty must nonetheless be put into perspective as shown in the following section.

8.4.3 Coordinates of the Factors of the Separate Analyses

The first two factors of the MFA (see Figure 8.1) are quite close to the factors of the same rank in the separate analyses. They are slightly more closely correlated with the factors of the quantitative group. This is probably a consequence of the low decrease in the inertias of the qualitative group, which indirectly favours the factors of the quantitative group in the sense that they mediate between directions of the qualitative group with comparable inertias.

Conclusion: Weighting the MFA worked very well in this example by balancing the influences of these two types of variable.

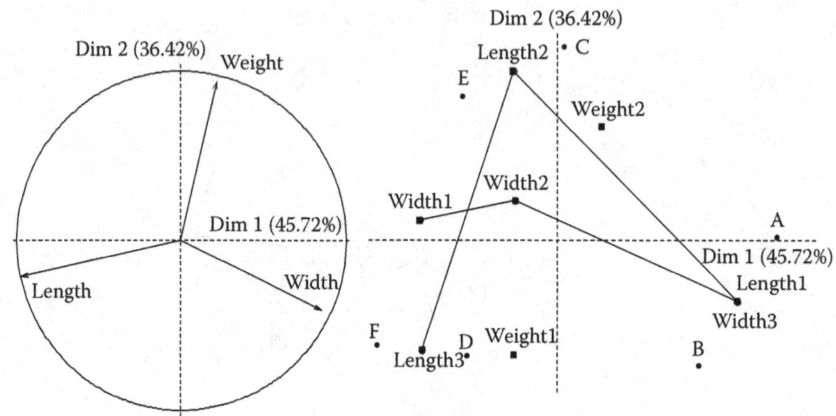

FIGURE 8.2
Biometry2. MFA. First plane. Left: Quantitative variables. Right: Individuals and categories.

This remark attenuates the impact of the reticence mentioned in the previous section concerning the simultaneous analysis of these two types of variable. Nevertheless, this also highlights the need to bear in mind the inertias of separate analyses when interpreting results of MFA. This necessity, which applies to any multiple table analysis, is exacerbated in cases of active groups of different types.

8.4.4 First Factor

Individuals and Variables (See Figure 8.2)
For quantitative variables, categories and individuals, the rules for interpreting the analyses are the same as for PCA and MCA.

The first factor:

- Is negatively correlated with length and positively correlated with width (quantitative)
- Ranks the individuals in almost regular (inverse) alphabetic order
- Ranks the categories of the qualitative variables *length* and *width* in their natural order, according to the previous correlations

In particular, this factor opposes individual *A*, wide and short, with *F*, long and narrow. This is indeed a factor common to the two groups of variables, with this opposition being clear in the data for each of these two types of variables.

Partial Individuals (See Figure 8.3)
There are some marginal differences between the partial individuals of the two groups. For axis 1, individuals *A* and *B* are thus more different from

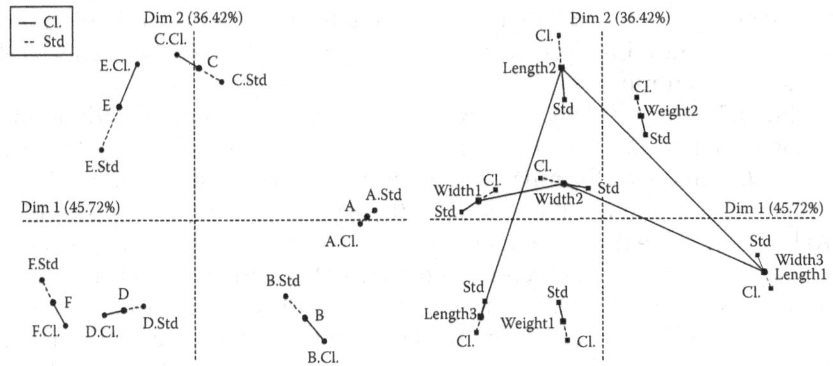

FIGURE 8.3
Biometry2. MFA. First plane. Individuals, mean and partial (left); Categories, mean and partial (right). Std: Standardised; Cl: Classes.

the point of view of the quantitative group (group 2 = Std) than from the point of view of the qualitative group (group 1 = Cl): $\left|F_1\left(A^1\right)-F_1\left(B^1\right)\right| < \left|F_1\left(A^2\right)-F_1\left(B^2\right)\right|$. Here we can see a consequence of coding the data into classes which can group together (slightly) different individuals: A and B differ (slightly) in terms of length and width, but this is not visible when grouped into classes (only their differences in weight, a variable very slightly linked to the first factor, prevent A^1 and B^1 from having the same coordinate on this axis).

Our attention is drawn to individuals D and E. In terms of axis 1, E is more typical than D if we consider the quantitative data: $\left|F_1\left(E^2\right)\right| > \left|F_1\left(D^2\right)\right|$. Indeed, D is slightly longer than it is wide, whereas E is much longer than it is wide; this can be seen in the raw data, with these two variables having the same standard deviation.

However, still from the perspective of the first axis, D is more typical than E if we consider the data grouped into classes: $\left|F_1\left(D^1\right)\right| > \left|F_1\left(E^1\right)\right|$. First, it can be seen that the coding into classes has erased the difference mentioned above, in the sense that the deviation between the two individuals is now of only one class both for length and for width. From this perspective, these two individuals are identical, as can be observed in the proximity between the barycentre (not represented) of $\{length = 3; width = 2\}$ for individual D and that of $\{length = 2; width = 1\}$ for individual E. The difference between the partial points stems from the weight, which is slightly related to the first factor and clearly opposes D and E (particularly in the data coded into classes).

Transition Relations

The transition relations expressing the coordinate of an individual in terms of those of the quantitative variables on the one hand and of those of the categories (see Section 8.3.2), underlie the interpretation of the projection of the cloud of individuals and variables and therefore the comments in the previous paragraph.

Table 8.7 shows details of how the coordinate of a partial individual is calculated from the data. For factor 1, it brings together the terms on the right in the first two formulae of Section 8.3.2, for which the sum is the coordinate of an individual (up to a coefficient, each term is the result of the product of a value from the data table by the coordinate of a variable or category). These values can therefore be seen as the variables' contributions to the coordinates of the individuals (for example, the coordinate of A^2 is worth $1.165+0.080+1.016 = 2.261$). In calculating this coordinate, length and width play the same role and weight is of very little consequence. Between the two groups, these contributions to the coordinates can be compared, as shown by the two examples below:

- The qualitative variable *length* does not separate A and B; compared to the quantitative variable *length*, it depicts (along F_1) both A less typical ($.927 < 1.165$) and B more typical ($.927 > .699$).
- Weight is much less closely linked to F_1 than length, which is translated by lower contributions in both codings. These contributions (see Figure 8.4, right) are slightly higher in this qualitative case, as a sign of a stronger relationship between F_1 and weight in the qualitative coding ($\eta^2\,(F_1, weight\ qlt.) = .097$) than in the quantitative data ($r^2\,(F_1, weight\ quant.) = .048$).

Partial Categories (see Figure 8.3)

Overall, the homologous partial categories (that is to say, those relating to the same category) are closer to one another than the homologous partial individuals are. This result is classical: out of those individuals with the same category, opposite deviations between partial images cancel each other out when calculating barycentres. Example: *width* = 3 (shared by A and B) along F_1. Counter-example: *width* = 1 (shared by E and F, both of which are more extreme from the quantitative point of view) along F_1.

Conclusion

The negative correlation between length and width, the principal structure of the data insomuch as it concerns two out of three variables, can be clearly observed through both codings. The loss of information due to the coding into a qualitative variable leads to a slightly different appearance of the cloud of individuals, in particular:

- Regular distribution of individuals for the raw data
- Distribution which suggests two classes ($\{A, B\}$ and $\{C, D, E, F\}$) after coding into classes

The superimposed representation of the MFA works in quantitative, qualitative and mixed cases. Its interpretation is essentially based on the partial

TABLE 8.7
Biometry2. MFA. Contributions of the Table's Cells to the Partial Individuals' Coordinates[a]

		A	*B*	*C*	*D*	*E*	*F*
	Length	1.165	0.699	0.233	−0.699	−0.233	−1.165
Group 2	Weight	0.08	−0.16	0.32	−0.16	0.08	−0.16
(Quantitative)	Width	1.016	0.61	−0.203	0.203	−1.016	−0.61
	Partial Indiv. G2	2.261	1.148	0.35	−0.656	−1.169	−1.935
Group 1	Length=1	0.927	0.927	0	0	0	0
	Length=2	0	0	−0.228	0	−0.228	0
	Length=3	0	0	−0.228	0	−0.228	−699
	Weight=1	0	−0.225	0	−0.225	0	−0.225
	Weight=2	0.225	0	0.225	0	0.225	0
(Qualitative)	Width=1	0	0	0	0	−0.71	−0.71
	Width=2	0	0	−0.217	−0.217	0	0
	Width=3	0.927	0.927	0	0	0	0
	Partial Indiv. G1	2.079	1.63	−0.221	−1.14	−0.714	−1.633

[a] (Partial Indiv. G1 and G2) on the first factor.

transition relations, which are almost identical to those of the PCA and the MCA.

8.4.5 Second Factor

This factor:

- Essentially opposes individuals E and C (contribution: 59.8%) with the others
- Is positively correlated with the quantitative variable *weight* ($R^2 = .88$)
- Is linked to the qualitative variables *weight* ($\eta^2 = .80$) and *length* ($\eta^2 = .902$)
  These two coefficients are read in the relationship square (see Figure 8.4) commented later in the chapter.

According to this factor, E and C are characterised by a heavy weight **and** an intermediary length (which can easily be seen in the raw data).

This second factor is common to both groups (the opposition between E and C with the others is clear in the two types of variables) but does not have exactly the same meaning in the two cases. It is linked to both *weight* variables but its meaning is enriched, for group 1, by the opposition between *extreme lengths* and *intermediary lengths*.

Here we notice that MCA can highlight nonlinear relationships. The quantitative perspective only indicates that the second factor is not correlated with length and width. The qualitative perspective also depicts an absence of relationship with width, but a nonlinear relationship with length.

From the point of view of the superimposed representation of partial clouds, it should be noted that the dispersion of partial individuals is greater for group 1 than for group 2. This is a consequence of the lower decrease of inertias of the separate analysis of group 1. In parallel, factor F_2 of the MFA is more closely linked to group 1 than group 2: $Lg\,(F_2, K_1) = .91$; $Lg\,(F_2, K_2) = .62$. As a result, F_2 separates the partial categories of group 1 more markedly.

8.4.6 Third Factor

This factor is specific to group 1 and closely correlated to the third factor of the MCA for this group ($r\left(F_3^{AFM}, F_3^1\right) = .95$). Being three-dimensional, group 1 could not be perfectly represented on the first plane of the MFA.

In MCA, each variable generates, in $\mathbb{R}^I$, a subspace of dimension (number of categories −1). For group 1, the first plane of the MFA takes into account the two length dimensions, the single weight dimension and one width dimension. Thus, the third factor of the MFA 'automatically' expresses the remaining width dimension.

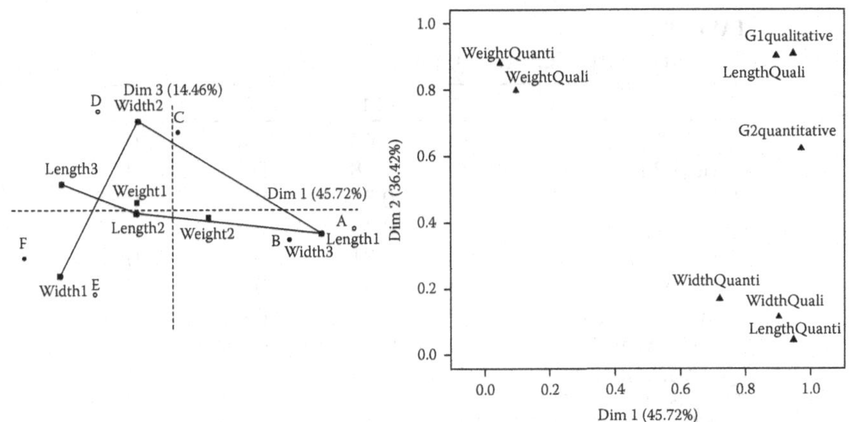

FIGURE 8.4
Biometry2. MFA. Left, individuals and categories on the ($F1$, $F3$) plane. Right, relationship square ($F1$, $F2$).

This third factor (see Figure 8.4) opposes categories 1 and 2 of width with each other (these categories are close on the first plane, particularly from the point of view of group 1, which corresponds to the identical way in which they associate themselves with the other categories) and, in parallel, the little wide individuals (E , F) and the moderately wide individuals (D, C).

8.4.7 Representing Groups of Variables

The graphical representation of groups of variables in the relationship square is particularly useful when there are many groups, which is not the case in this example. However, this representation can be completed by those of the variables, each considered as a group (the coordinate of a group made up of a single variable is a squared correlation coefficient/ratio).

Figure 8.4 (right) provides a synthetic visualisation of the two main outlines of the interpretation which has already been discussed:

1. The first factor of the MFA corresponds to a high direction of inertia for the two groups of variables; it is strongly linked to two out of three variables of each (length and width).
2. The second factor of the MFA corresponds to a high direction of inertia for the first group (it is related to two variables of this group: weight and length) and less for the second group (it is related only to weight).

As the groups of variables are not normed, their qualities of representation (in the sense of *projected inertia/total inertia*) do not appear on the graph and must be consulted separately (see Table 8.8). Thus, on the first plane:

TABLE 8.8
Biometry2. MFA. Groups' Quality of Representation[a]

	F1	F2	F3
Group 1 (Classes)	0.43	0.40	0.14
Group 2 (Standardised)	0.68	0.28	0.00
All	0.53	0.35	0.08
Quali. length	0.40	0.41	0.00
Quali. weight	0.01	0.64	0.00
Quali. width	0.41	0.01	0.48
Quanti. length	0.90	0.00	0.00
Quanti. weight	0.00	0.78	0.00
Quanti. width	0.52	0.03	0.01

[a] In $\mathbb{R}^{I^2}$ on the first three axes.

- The distance between the two codings of the length (well represented) do indeed correspond to unidimensionality of one and bidimensionality of the other.
- The proximity between the two widths (poorly represented) does not correspond to an overall equivalence between the two codings, which is impossible due to the different dimensions; the difference between the two codings is expressed on axis 3.
- The proximity between the two weights (quite well represented) suggests an equivalence between the two codings, possible from the point of view of the dimensions and not contradicted by the other axes.

The influence of coding can be measured by the relationship indicators Lg and RV calculated between the two codings of the same variable (see Table 8.9). As the groups are reduced to a quantitative variable and a qualitative variable, Lg is interpreted as the squared correlation ratio. As the qualitative variable *weight* has two categories, in this case Lg is interpreted as the squared correlation coefficient between the quantitative variable and one of the two indicators. Finally, in this very specific case, RV is equal to Lg divided by the square root of the number of categories of the qualitative variable.

Indicator Lg identifies a weaker common structure between the two codings for the weight: the qualitative variable better reproduces the raw data in the case of length and width. The fact that these two variables have more categories mechanically increases Lg in this case. In comparison, the RV indicator brings into perspective the importance of the common structure through the number of categories of the qualitative variable: it indicates that the images of the individuals given by the two codings are closer for weight. These two points of view are not contradictory: the qualitative coding with more than two categories makes it possible to highlight nonlinear relationships, which are invisible with quantitative variables (in factorial analysis).

TABLE 8.9
Biometry2. Relationship Indicators Lg and RV[a]

Measurement/Variable	Length	Weight	Width
Lg	.914	.800	.914
RV	.646	.800	.646

[a] Between the two codings of each variable.

8.4.8 Conclusion

This example shows that the weighting of the variables included in MFA makes it possible to analyse groups of variables of different types simultaneously. The interpretations are based on both PCA and MCA. They do not pose any particular problem aside from the need to bear in mind the inertias of the separate analyses (classic in MFA but heightened in this type of application).

The advantage of studying quantitative variables by coding them as qualitative and then conducting an MCA has been confirmed (if ever confirmation was needed). The linear relationships detected by the PCA are also identified by the MCA. But MCA identifies other relationships. In this example, in which the relationships are essentially linear, the first factors of the PCA and the MCA are similar, but of course this is not always the case.

Here, MFA proves a rich and useful tool for comparing two methodologies. The advantage of this method when the groups have a lot in common is always observed.

Another application is suggested by this example: MFA makes it possible not to choose between two (or more) codings but rather to conduct an analysis which accounts for both. This approach will be valuable particularly to construct a classification of individuals from the coordinates of the MFA (which here plays the role of preprocessing) which is robust in terms of coding.

8.5 MFA of Mixed Data in FactoMineR

We do not go into detail with regard to the **R Commander** interface, which has already been described in Section 1.11, and instead focus on the specifics of the qualitative variables.

This section is illustrated with the Biometry2 data (see Table 8.3). Each variable is present in the file twice in an order which appears in the importation verification:

```
> Biometry2=read.table("Biometry2.csv",header=TRUE,sep=";",
+ row.names=1)

> colnames(Biometry2)
```

```
 [1] "Length3classes" "Weight2classes" "Width3classes"  "Length"
 [5] "Weight"         "Width"          "Length3cl."     "Weight2cl."
 [9] "Width3cl."      "Length2"        "Weight2"        "Width2"
```

The MFA is conducted on the first six columns only, with all of the default options (the names of the groups are thus `group.1` and `group.2`). The columns of each group must be consecutive in the table.

```
> res=MFA(Biometry2[,1:6],group=c(3,3),type=c("n","s"))
```

This command displays the main graphs: mean individuals, quantitative variables, mean and partial categories, partial axes, groups, mean and partial individuals (the latter being limited to four individuals with the two highest and the two lowest within inertias for the first axis).

For some graphs, it can be useful to have short group names. Thus, to label partial individuals and/or categories:

```
> res=MFA(Biometry2[,1:6],group=c(3,3),type=c("n","s"),
+ name.group=c("Class.","Std"))
> plot.MFA(res,axes=c(1,2),choix="ind",habillage="group",
+ invisible="quali",partial="all",lab.par=TRUE)
```

The graph in Figure 8.3 on the right is obtained as follows:

```
> plot.MFA(res,axes=c(1,2),choix="ind",habillage="group",
+ invisible="ind",partial="all")
# Labelling partial points with group labels
> text(res$quali.var$coord.partiel[seq(1,15,2),1],
+ res$quali.var$coord.partiel[seq(1,15,2),2],
+ rep("Std",6),pos=3,offset=0.5)

> text(res$quali.var$coord.partiel[seq(2,16,2),1],
+ res$quali.var$coord.partiel[seq(2,16,2),2],
+ rep("Cl.",6),pos=1,offset=0.5)

# Connecting categories of variables length and width
> varco<-res$quali.var$coord
> points(varco[1:3,1],varco[1:3,2],type="o")
> points(varco[6:8,1],varco[6:8,2],type="o")
```

It is often useful to colour the individuals according to the categories they possess for a specific qualitative variable. Thus, to differentiate between the individuals according to the weight (`hab=2`):

```
> plot.MFA(res,choix="ind",invisible="quali",hab=2)
```

It is possible to display a confidence ellipse around the mean point of the individuals possessing a given category of a given variable, using the `plotellipses` function. For example, for the categories of the weight (`keepvar = 2`):

```
> plotellipses(res,keepvar=2)
```

The relationship square (see Figure 8.4, right) can display the variables themselves as well as the groups of variables. In order to do this, each variable is also introduced as a supplementary group of one single variable. This is why the variables are duplicated in the file. In this graph, longer labels are used for the groups. Thus:

```
> res=MFA(Biometry2,group=c(3,3,rep(1,6)),
+ type=c("n","s",rep("n",3),rep("s",3)),
+ num.group.sup=c(3:8),name.group=c("G1qualitative",
+ "G2quantitative","LengthQuali","WeightQuali","WidthQuali",
+ "LengthQuanti","WeightQuanti","WidthQuanti"))
```

Among other things, this command generates the relationship square seen in Figure 8.4.

Most of the tables in this chapter are not specific to mixed data and the way in which they are obtained has already been described. The four tables below are (more or less) new.

Table 8.5 (Correlations Between Partial Axes)
The program brings together the correlation coefficients for all pairs of factors in a table, from which we extract a section:

```
> round(res$partial.axes$cor.between[6:8,1:3],2)
```

Table 8.7 (Contributions of the Cells to Partial Individuals)
This table is very particular and, unlike the other three, it is seldom constructed. It is important to appreciate subtly how the quantitative and qualitative variables are balanced. It is mainly based on the partial transition relations described in 8.3.2.

The `tab.disjonctif` function is included in **FactoMineR**. From a data.frame containing the qualitative variables (factors), it constructs a complete disjunctive table (CDT). This function has many other uses, particularly for grouping together contingency tables crossing subsets of qualitative variables with one another in one single table.

```
# Initialisation
> Tab8_7=matrix(rep(0,78),nrow=13,ncol=6)
# BCR: Quantitative group standardised (centred and reduced)
# Bdis: qualitative group (complete disjunctive coding)
> BCR=as.matrix(scale(Biometry2[,4:6])*sqrt(6/5))
> Bdis=tab.disjonctif(Biometry2[,1:3])
> colnames(Tab8_7)=rownames(Biometry2)
> rownames(Tab8_7)=c(colnames(BCR),"Ind.part.quanti",
+ colnames(Bdis),"Ind.part.quali")

# First eigenvalue of MFA and of separate analyses
> L1AFM=res$eig[1,1]
> L1ACM=res$separate.analyses$Categ.$eig[1,1]
> L1ACP=res$separate.analyses$Std$eig[1,1]
```

In the transition relation in PCA (for the axis of rank s), a coefficient equal to the root of the eigenvalue (of rank s) appears; the role of this coefficient is here held by the first eigenvalue of the MFA (first as here we are interested in the axis of rank 1). In addition, the weighting of the MFA by the first (first as it is the choice of the weighting of the MFA) eigenvalue of the separate PCA for the groups of quantitative variables intervenes here.

```
# Coeff = coefficient in transition formula
> coord=res$quanti.var$coord[,1]
> coeff=2/(sqrt(L1AFM)*L1ACP)
```

Index j (columns) is that of the individuals (there are six) and index i (rows) that of the variables (compared with the usual format, the table is transposed to make it easier to edit):

```
# Quantitative group part
> for(i in 1:3){ for (j in 1:6) {
+ Tab8_7[i,j]=BCR[j,i]*coord[i]*coeff
+ Tab8_7[4,j]=Tab8_7[4,j]+Tab8_7[i,j]
+ }}
```

For qualitative variables, the first eigenvalue of the PCA is replaced by the first eigenvalue of the MCA multiplied by the number of variables. In addition, the first eigenvalue of the MFA here intervenes directly, rather than through its root, as here we use the coordinates of the centres of gravity for the categories and not those of the indicators (see Section 8.1.2).

```
# Qualitative group part
> coeff=2/(L1AFM*L1ACM*3)
> coord=res$quali.var$coord[,1]
> for(i in 1:8){ for (j in 1:6) {
+ Tab8_7[i+4,j]=Bdis[j,i]*coord[i]*coeff
+ Tab8_7[13,j]=Tab8_7[13,j]+Tab8_7[i+4,j]
+ }}
```

Table 8.8 (Qualities of Representation in $\mathbb{R}^{I^2}$)
To make editing easier, slightly shortened labels are used.

```
# Initialisation and choice of labels
> tab_8.8=matrix(nrow=9,ncol=3)
> row.names(tab_8.8)=c("Group 1","Group 2","Overall",
  "LengthQuali",
+ "WeightQuali","WidthQuali","LengthQuanti","WeightQuanti",
+ "WidthQuanti")
> colnames(tab_8.8)=c("F1","F2","F3")

# The cos2 of Wj are calculated by MFA (cos2 and cos2.sup)
```

```
> tab8_8[1:2,1:3]=res$group$cos2[,1:3]
> tab8_8[4:9,1:3]=res$group$cos2.sup[,1:3]

# Projected inertia/global inertia ratio (for NJ)
# must be calculated from coordinates (coord)
# and distances between the Wj and the origin (dist2)
> tab8_8[3,1:3]=apply(res$group$coord[,1:3]^2,MARGIN=2,FUN=sum)
> tab8_8[3,1:3]=tab8_8[3,1:3]/sum(res$group$dist2)

# Editing numbers with 2 decimal places
> round(tab8_8,2)
```

Table 8.9 (*Lg* and *RV* Relationship Indicators)
To obtain this table, as for the relationship square, each variable was also introduced as a supplementary group of just one variable. The desired *Lg* and *RV* indicators are found in the matrices bringing together these coefficients for each pair of groups (`res$group$Lg` and `res$group$RV`):

```
# Initialisation and choice of labels
> tab8_9=matrix(nrow=2,ncol=3)
> row.names(tab8_9)=c("Lg","RV")
> colnames(tab8_9)=c("Length","Weight","Width")

# Lg and RV coefficients are on the diagonal of submatrices
# included in res$group$Lg and res$group$RV
> tab8_9[1,1:3]=diag(res$group$Lg[3:5,6:8])
> tab8_9[2,1:3]=diag(res$group$RV[3:5,6:8])

# Editing numbers with 3 places
> round(tab8_9,3)
```

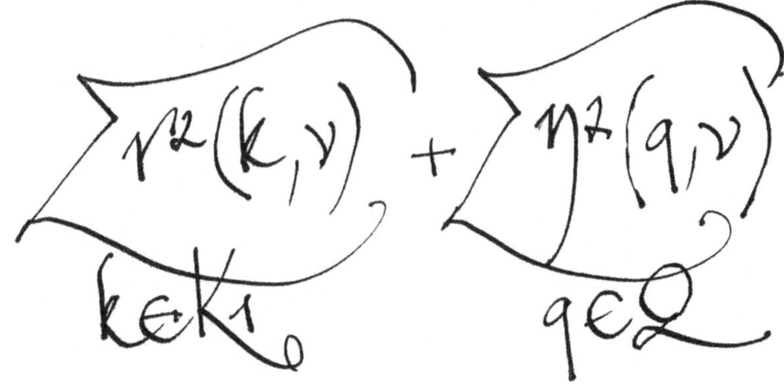

9

Multiple Factor Analysis and Procrustes Analysis

Originally, the question which motivated Procrustes analysis (PA) was, 'How can one of two homologous clouds of points situated within the same space be turned to make them coincide as closely as possible?' Today, generalised Procrustes analysis (GPA) is applied to sets of more than two clouds.

Chapter 5 describes the solution proposed by multiple factor analysis (MFA) to obtain a superimposed representation of homologous clouds of points (N_I^j). The two approaches must be compared.

As GPA is not well known outside the field of sensory analysis, we start by presenting some basic elements of this method.

9.1 Procrustes Analysis

9.1.1 Data, Notations

There are J clouds (denoted N_I^j) of I homologous points (i^j). These clouds evolve within spaces of the same number of dimensions K_c (but these dimensions do not correspond from one space to the other). The coordinates of the points of N_I^j are brought together in matrix X_j of dimensions (I, K_c). The columns of X_j are centred.

When initially N_I^j evolve within spaces of different dimensions (K_j), we choose $K_c = \max(K_j, j = 1, J)$; when $K_j < K_c$, we consider that N_I^j has zero inertia in $K_c - K_j$ directions, which it obtained by adding columns of 0 to the initial matrix X_j.

In Procrustes analysis, the individuals are always attributed a weight of 1. These weights are used in this chapter. That being said, the presence of different weights from one individual to the other is not contradictory to Procrustes analysis: we might want the superimposition to be a particularly good fit for certain individuals. Reminder: in MFA it is possible to attribute weights to individuals, a possibility which is offered by **FactoMineR**.

9.1.2 Objectives

Clouds N_I^j are positioned in one space $\mathbb{R}^{K_c}$. Each cloud is centred and no translation is required. We then transform N_I^j so as to make the homologous points coincide as well as possible. In the most common (and the original) version, only orthogonal transformations are authorised (that is to say, the rotations and symmetries) as they do not modify the distances between the points of a single cloud. Homotheties can also be authorised, but their application advantage cannot be seen when N_I^j have been standardised. Thus, unless explicitly stated to the contrary, they are not considered.

Following transformations, cloud N_I^j has new coordinates that are brought together in matrix Y_j. Saying that N_I^j coincide (with one another) as well as possible is the same as saying Y_j are as close together as possible. The sum of the square of the term-by-term differences between matrices Y_j and Y_l is written:

$$\text{trace}\left[(Y_j - Y_l)'(Y_j - Y_l)\right].$$

The quantity which Procrustes analysis minimises is thus written:

$$\sum_{j>l} \text{trace}\ (Y_j - Y_l)'(Y_j - Y_l).$$

Y_j is deduced from X_j by isometry (a transformation which preserves the distances); Y_j can be written $Y_j = X_j T_j$ with T_j as an orthogonal matrix, that is to say, verifying $T_j T_j' = I_d$ (denoting I_d the identity matrix of the appropriate size). The Procrustes model can therefore be expressed:

$$q_j X_j T_j = Z + E_j,$$

where Z, matrix of size (I, K_c), contains the coordinates of the so-called mean configuration (in reference to the way it is calculated), E_j a matrix of residuals and q_j a scalar present in the model when the homotheties are authorised.

Mean Cloud. When N_I^j are placed in the same space, we can construct a mean cloud N_I for which each point i is the isobarycentre of its homologous points in N_I^j. Up to a certain point, this cloud is analogous with mean cloud N_I of the MFA. To make it easier to compare the two methods, we call N_I the mean cloud, whatever the analysis (even though the two clouds are not constructed in the same way).

9.1.3 Methods and Variations

Depending on the Number of Clouds

First Case: $J = 2$. This is the original method. It includes an analytical solution which we look at briefly.

Let X_1 and X_2 be the tables containing the initial data which are generally centred and reduced; we aim to transform X_2 to fit X_1.

Let $V_{12} = X_1' X_2$ be the matrix containing (up to I coefficient) the correlation coefficients between the variables of group 1 (rows) and those of group 2 (columns), when the data are centred and reduced. V_{12} contains the covariances if the data are only centred.

Let U be an (orthogonal) matrix of standardised eigenvectors of $V_{12} V_{12}'$ and V an (orthogonal) matrix of standardised eigenvectors of $V_{12}' V_{12}$.

It can be shown that the fit of table X_2 (to table X_1) is given by

$$Y_2 = X_2 V U'.$$

The dissymmetry of the solution is only apparent: from the point of view of the relative position of the points, there is no difference between fitting X_1 to X_2 or X_2 to X_1.

Remark

The axes of the principal component analysis (PCA) only depend on the correlations between variables; those of the Procrustes analysis depend only on between-table correlations.

Second Case: $J > 2$. This is the case of the GPA. There is no known analytical solution. An iterative algorithm, at each step, successively fits each cloud N_I^j to the mean cloud (in the first step, the first cloud acts as a mean cloud). The mean cloud itself is recalculated after the rotations of N_I^j. More specifically, the principle of the usual algorithm can be described as follows:

1. Initialise the mean cloud Z (by the first configuration).
2. Fit each of J clouds N_I^j to the mean cloud; update N_I^j by the result of these fits which are conducted successively.
3. Update the mean cloud Z from J fitted clouds N_I^j.
4. Update the fitting criterion for all N_I^j.
5. Begin step 2 again while the improvement of the criterion is above a fixed threshold.

This algorithm converges, but its convergence to an overall optimum of the criterion is uncertain. Many studies have tried to improve this algorithm.

Depending on the Number of Dimensions

First Case: $K_c = 2$ **or** 3. The solution can be examined directly and overall by a graphical representation.

Second Case: $K_c > 3$. The solution can only be examined via a projection on subspaces. In the usual variant, at the end of the PA (or the GPA), N_I^j are projected onto the factorial axes of N_I. It has also been suggested to use the factorial axes of the union of the N_I^j (denoted N_I^j).

Influence of the Dimensions on the Objectives
When $K_c > 3$, homology between the spaces in which N_I^j evolve is not studied globally but subspace by subspace. In practice, the representations are often studied dimension by dimension. This point of view brings GPA closer to generalised canonical analysis, that is to say, the search for a sequence of directions common to several homologous clouds of points. We often refer back to this approach.

9.2 Comparing MFA and GPA

Reminder: The MFA is based on a weighted PCA of table X of size (I, K) juxtaposing tables X_j in rows. In this PCA, the variables of group j are weighted by $1/\lambda_1^j$ (denoting λ_1^j the first eigenvalue of the separate PCA for group j).

9.2.1 Representing N_I^j

Each cloud N_I^j corresponds to a table X_j.

In MFA
N_I^j are placed in space $\mathbb{R}^K$, the direct sum of $\mathbb{R}^{K_j}$. Thus, N_I^j are not really in the same space. The simultaneous nature of this representation is artificial; it is justified as a framework for interpreting the method.

In GPA
N_I^j are all positioned in the same space $\mathbb{R}^{K_c}$. This representation corresponds to an overall homology of spaces $\mathbb{R}^{K_c}$ which each contain a cloud N_I^j.

N.B.: Initially, it is an overall homology, rather than initial dimension by initial dimension homology, as is the case when the variables are the same from one group to the other. This initial superimposed representation of N_I^j is artificial, as is that of the MFA, and is justified as a framework for interpreting the method.

The aim of GPA is to identify the homologous dimensions of N_I^j from this overall homology of spaces $\mathbb{R}^{K_c}$. These homologous dimensions induce the same structure on the individuals. We here observe the common factor notion of MFA.

Almost all of the differences between the two methods are derived from the difference between the two modes of representation of N_I^j.

9.2.2 Mean Cloud

In both methods, mean cloud N_I contains the points i, the centres of gravity of the sets $\{i^j; j = 1, J\}$ (which is why they are called mean clouds). However, as these two mean clouds N_I are constructed in different spaces, they do not carry the same meaning from one method to another.

In MFA

- The i^j associated with a given i belong to orthogonal subspaces.
- Their coordinates are juxtaposed and cannot be averaged.
- The squared distance between two mean points i and l is written:

$$d^2(i,l) = \frac{1}{J^2} \sum_j d^2(i^j, l^j).$$

- This overall distance increases with the distances in each subspace, whatever the direction of this distance in each subspace.
- In $\mathbb{R}^K$, the total inertia of N_I is equal to the total inertia of N_I^J (union of N_I^j) divided by J^2.

In GPA

- The i^j associated with a given i belong to the same space, $\mathbb{R}^{K_c}$.
- The coordinates of i are the average of the coordinates of $\{i^j; j = 1, J\}$.
- As in MFA, the distance between the two mean points i and l is dependent on the distance between individuals i and l in each group but, in addition, on the fact that, along homologous directions, the deviations between individuals i and l either do or do not have the same sign (see Figure 9.1).
- As a result, at constant inertia N_I^J, the inertia of N_I is greater when the deviations between the points are identical in homologous directions, that is to say, when N_I^j are similar. This is why, in GPA, the mean cloud is sometimes called a consensus cloud.

Illustration. In Figure 9.1, A and B differ for each group: in MFA the mean points differ. The same applies for C and D. On the axis of the PA, the differences between A and B are similar for each group: the mean points differ. On this axis, the differences between C and D are opposite for the two groups: the mean points are the same.

9.2.3 Objective, Criterion, Algorithm

From a general point of view, in both cases we are looking for factors common to N_I^j which are visualised using a superimposed representation of N_I^j.

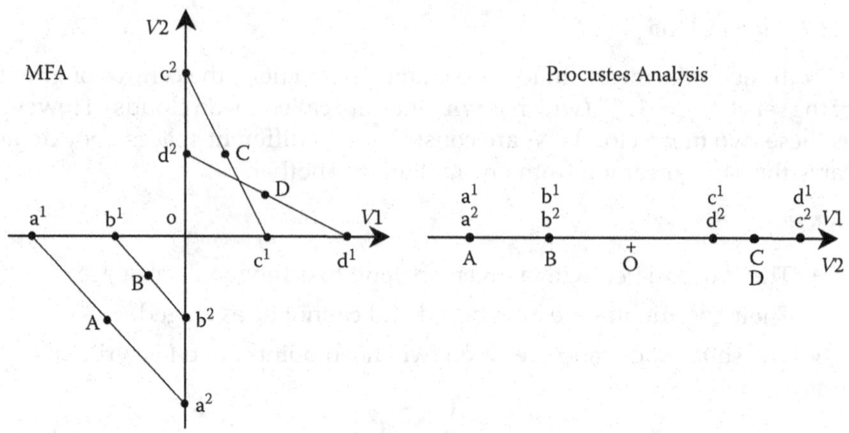

FIGURE 9.1
Four individuals (A, B, C, D) described by two groups, each composed of one variable ($\{V1\}$, $\{V2\}$). Representations of clouds N_I and N_I^j in space $\mathbb{R}^K$ of the MFA (left) and on the unique axis of the Procrustes analysis (right).

In MFA

- The number of common factors is not specified.
- The common factors can be common to all or some of the groups.
- The aim of the analysis is to highlight them.

In GPA

- We assume that there is an homology between the spaces in which N_I^j initially evolve. In practice, we identify homologous bases of these spaces. Finally, in terms of canonical analysis, this means assuming that there are K_c factors common to all of the J groups.
- The aim of the analysis is to identify these common factors.

Criterion
In both methods, the criterion can be expressed from the superimposed representation. We consider the partition of N_I^j into I classes, each containing the partial points associated with one given individual $\{i^j; j = 1, J\}$. The within-class inertia of N_I^j relative to this partition is written:

$$\sum_i \sum_j d^2\left(i^j, i\right) = \frac{1}{2I} \sum_i \sum_{j,l} d^2\left(i^j, i^l\right).$$

In PA and GPA, the aim is to minimise this quantity. When homotheties are not authorised, the same result can be achieved by maximising the associated between-class inertia which is none other than the inertia of N_I. For GPA, at each step we are looking for the rotation of each cloud N_I^j which satisfies this criterion (thus calculated for all dimensions).

Remark

When homotheties are authorised, one trivial solution is to choose 0 as the q_j coefficient for each cloud j. In order to avoid this, one can fix the total inertia of N_I^j (union of N_I^j).

In MFA, axis by axis, the aim is to maximise projected between-class inertia. Despite their resemblance, the two criteria only partially correspond because, by proceeding axis by axis, total inertia is no longer fixed, which cancels out the equivalence of minimising within-inertia and maximising between-inertia.

Therefore, between the two methods:

- The geometric representations differ; the mean clouds do not have exactly the same meaning.
- The quantities to be maximised differ despite the existence of a resemblance.
- The types of transformation of N_I^j differ (rotation or projection).

9.2.4 Properties of the Representations of N_I^j

In GPA

- The transformations of N_I^j are orthogonal (which may include homotheties).
- The shape of N_I^j is perfectly respected; this is a very strong constraint, specific to GPA.
- If K_c is greater than 3, the superimposed representation of N_I^j can only be examined using projections, for example, on the principal axes of N_I; these projections are conducted after fitting.

In MFA

- The projection is conducted in parallel to the fitting.
- The projection of N_I^j is conducted on axes which do not belong to $\mathbb{R}^{K_j}$ or, from another point of view, on nonorthogonal axes of $\mathbb{R}^{K_j}$. The result is a deformation of N_I^j, even if cloud N_I is perfectly represented. These distortions have already been described in Section 5.5; an illustration is given in the example (referred to as 2^{3-1}) described below in Section 9.3.

9.2.5 A First Appraisal

MFA is a particular factorial analysis and a particular multicanonical analysis (in Carroll's sense). It is not a Procrustes analysis if we consider the nondeformation of N_I^j as one of its characteristics. Nonetheless the issues which arise in Procrustes analyses and MFA are connected:

TABLE 9.1
Three Individuals (A, B, C) Described by Three Groups[a]

	$V1$	$V2$	$V3$		$V1$	$V2$	$V3$
A	5	1	−3	$V1$	1		
B	−2	−2	−2	$V2$	.40	1	
C	−3	1	5	$V3$	−.68	.40	1
		Data				Correlations	

[a] Each with a single variable.

- Both include the notion of the common factor.
- Both include a superimposed representation of N_I^j and a mean cloud.

9.2.6 Harmonising the Inertia of N_I^j

In MFA, the inertia of N_I^j is harmonised prior to the analysis:

- Within the groups, by optional reduction of the columns
- Between the groups, by overweighting the variables which amounts to a homothety of each N_I^j

In GPA, the issue is the same as prior to conducting an MFA or any other analysis of this type of multiple table. The practice is: the inertia of N_I^j is always 100; optionally, the inertia of each N_I^j can also be fixed to $100/J$.

9.2.7 Relationships Between Homologous Factors

We use the notations from MFA: F_s: coordinates of N_I along axis of ranks; F_s^j: coordinates of N_I^j along axis of rank s.

It can be shown then, in MFA, these factors benefit from the following property:

$$\forall s, j : r\left(F_s^j, F_s\right) \geq 0.$$

Thus, a canonical variable F_s^j is never negatively related to the general variable of the same rank F_s. This property is the very least we can ask of a common factor. The GPA can also be shown to verify this property.

However, in both GPA and MFA, homologous factors (such as F_s^j and F_s^l) can be negatively correlated, as illustrated in the example (see Table 9.1).

In this specific case, where each group has only one single variable, the GPA considers the three variables to be homologous (sometimes considering their opposites). Thus, the correlation matrix between homologous factors is the same as the correlation matrix between initial variables (if necessary accounting for symmetries by changing the signs of one or more rows and their corresponding columns). When a variable is positively correlated with two variables which themselves are negatively correlated with each other (as is the case here), there are negatively correlated homologous factors.

Remark
If, in the data from the previous example, GPA are conducted using homotheties, group 2 is attributed a 0 coefficient, which illustrates how introducing homotheties makes it possible to exclude a group. However, introducing homotheties satisfactorily solves the problem of negative correlations between homologous dimensions only in one-dimensional cases, as the homothety is applied in the same way to all of the dimensions of the group.

9.2.8 Representing Individuals

Preliminary Note. When the Procrustes model is verified exactly (that is to say, when N_I^j infer one another by rotation or symmetry), both methods provide the 'correct answer', that is to say, a superimposed representation of N_I^j in which the homologous points are equal and the shapes of N_I^j perfectly respected. For GPA, it is clear as the mean cloud is identical to each N_I^j after rotation. For MFA, consider three groups $X = (Z, ZA, ZB)$ with $A'A = I_d$ and $B'B = I_d$ (with I_d the identity matrix of the appropriate size); the principal components of the MFA are eigenvectors of $X' = 3Z'Z$: the mean cloud and each partial cloud therefore have the same principal components.

In identifying common factors, the GPA framework is restricted compared to that of MFA, as it supposes that:

- There are K_c common orthogonal factors.
- The factors are common to all groups.

These constraints weigh on all of the results (in particular on the first factors), as we are looking to find an overall optimum. Let us consider the case of a factor common to only certain groups.

- In MFA, N_I^j which do not possess this factor are orthogonal to it; they have no influence on it.
- In GPA, all directions are common to all the groups and a factor common to only some groups will be superimposed with the direction of the other groups with which it is not related. In this case, identifying this common factor can be disturbed: the mean configuration does not correspond to that of this common factor as it is distorted by the representations of the groups which have nothing in common with it, but which are nonetheless superimposed on it.

9.2.9 Interpretation Aids

We present the principal aids of GPA, specifying their meaning and, if appropriate, their equivalents in MFA.

In GPA, superimposed representation provides a framework in which the total inertia of N_I^J can be decomposed in many ways and can induce a comprehensive system of indicators (this decomposition is presented for MFA in Section 5.4). Total inertia (in practice set at 100) is first decomposed into between-inertia (inertia of the consensus N_I) and within-inertia (inertia of N_i^J clouds, each bringing together J partial points associated with individual i). These three inertias are then themselves decomposed in three ways (an example of these decompositions is given below in Section 9.3.3).

Decomposition by Dimension

- Between-inertia indicates the relative importance of the dimensions; here we find the eigenvalues of the PCA of N_I. This indicator is the same as for MFA.
- Within-inertia indicates the degree of consensus of the dimension. This indicator is the same as that in the MFA, in which we divide between-inertia to total inertia (see Section 5.4 and Table 5.2).

Decomposition by Group

- The proportion of group j in within-inertia measures the similarity between N_I^j and N_I. In MFA, we calculate the canonical correlation coefficients and the Lg measurements thus making it possible to evaluate the relationship between N_I^j and N_I axis by axis. There are two main differences between the methods: the nature of the indicator and whether they are calculated axis by axis or overall. These differences follow the different perspectives of the two methods. It is possible to introduce the indices of one method in the other, except for the sums of within-inertia on several axes which have no meaning in MFA.
- Decomposing within-inertia would make no sense here.

Decomposition by Individual

- The between-inertia of individual i is its contribution to the mean cloud; this indicator also exists in MFA (careful: the mean cloud does not have exactly the same meaning) but then it is mostly used axis by axis as in any factorial analysis.
- The within-inertia of individual i indicates whether this individual is the object of overall consensus. In MFA, this indicator is calculated axis by axis but cannot be cumulated over several dimensions; in GPA, it can be calculated overall and broken down axis by axis.

Summary

- Both methods have systems of indicators which make it possible to scan all of the issues that arise when examining a superimposed representation.
- However, in MFA it is not possible to add together the projected inertias of the N_I^j over several axes.
- In GPA, the forced nature of the superimposition, which weighs on the consensus, also consequently weighs on the indicators as shown in the following example (2^{3-1}).

9.2.10 Representing the Variables

In both methods, the correlation coefficients (along with the covariances where appropriate) are calculated between the initial variables and the dimensions of the mean cloud. These coefficients are represented graphically in the same way as in PCA.

Here, the difference between the two methods is caused by the fact that in MFA the initial variables play a direct active role (through the within- and between-group relationships) in representing the mean cloud, whereas in GPA they intervene indirectly. As a result:

- The representation of variables has its own optimality in MFA, which is not the case in GPA.
- The transition relation which expresses the coordinate of a mean individual according to the variables' coordinates does not exist in GPA.
- The relationship which in MFA expresses the coordinate of a partial individual (i^j) according to the coordinates of the variables (of group j) is all the more specific to MFA. This relationship, described in Section 5.2 (property 3), is key to the interpretations.

9.3 Application (Data 2^{3-1})

9.3.1 Data 2^{3-1}

Four individuals (A, B, C and D) are described by three groups, each with two variables. The data are presented in Table 9.2 and illustrated in Figures 9.2 and 9.3.

The six variables are centred. They derive from three variables X, Y and Z, which are uncorrelated pairwise, with a variance of 1, and which have been multiplied by 2, 3 or 6. These variables are therefore the principal components of the separate PCAs. The three variables X, Y and Z are constructed from

TABLE 9.2
Data 2^{3-1}

	Group 1		Group 2		Group 3	
	X_1	Y_1	X_2	Z_1	Y_3	Z_3
A	6	6	6	−2	3	−6
B	6	−6	6	2	−3	6
C	−6	6	−6	2	3	6
D	−6	−6	−6	−2	−3	−6
Variance	36	36	36	4	9	36

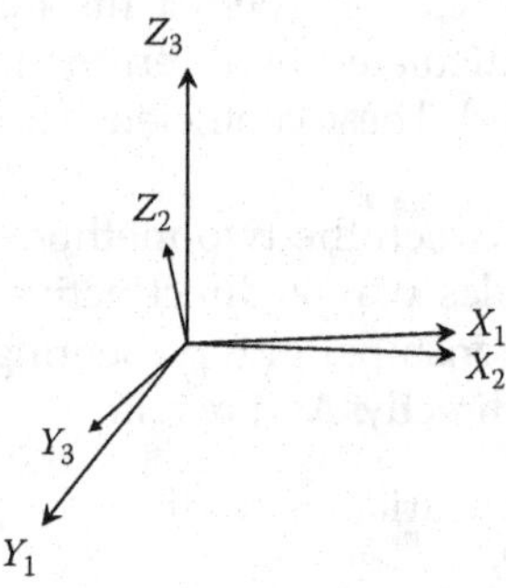

FIGURE 9.2
Data 2^{3-1}. Representation of the six variables in $\mathbb{R}^4$. As the variables are centred, they are situated in a three-dimensional subspace, which makes it possible to represent them. The correlation coefficients are here either 0 or 1. The perfectly correlated variables (example X_1 and X_2) are slightly separated in order to represent them distinctly.

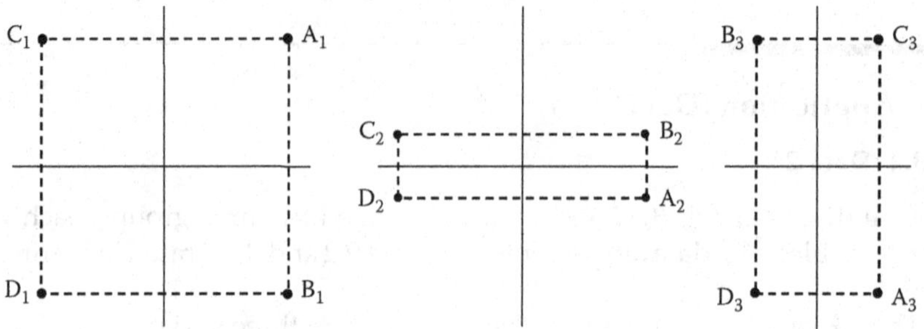

FIGURE 9.3
Data 2^{3-1}. Representation of the four individuals for each of the three groups.

TABLE 9.3
Data 2^{3-1}. Relationships Between Factors of the MFA and Groups

	F1	F2	F3	F1	F2	F3
Group 1	1	1	0	1	1	0
Group 2	1	0	1	1	0	1/9
Group 3	0	1	1	0	1/4	1
		$r(F_s, F_s^j)$			$Lg(F_s, j)$	

factors of the fractional factorial design 2^{3-1}, which explains the name of the data. The variables are not reduced, which leads to uneven directions of inertia. By proceeding in this way, each group presents a factor common to each of the other two; these common factors are not necessarily associated with the same inertia from one group to another.

The maximum inertia is the same in each group, which eliminates the influence of the weighting of the MFA: both analyses operate on the same data.

9.3.2 Results of the MFA

Projected Inertias of the Mean Cloud

The inertias of the axes of the MFA are easy to calculate. The first axis corresponds to variable X. Its inertia is obtained by adding together those of X_1 and X_2 (after weighting by the MFA, that is to say, by dividing each variance by the maximum variance of its group). Thus $36/36+36/36 = 2$. Axis 2 corresponds to variable Y and has an inertia of $36/36+9/36 = 1.25$. Axis 3 corresponds to Z with an inertia of $4/36 + 36/36 = 10/9$. In this analysis, three dimensions are needed to represent the data (data are generated from three orthogonal variables). They are of similar importance.

Relationship Measurements Between Factors and Groups

Two measurements are used and brought together in Table 9.3 in which we find, at the crossing of group j and factor s:

- The (canonical) correlation coefficient between F_s and F_s^j
- The relationship measurement Lg between F_s and group j

The canonical correlation coefficients indicate that $F1$ is common to groups 1 and 2, $F2$ to groups 1 and 3, $F3$ to groups 2 and 3. The Lg measurements specify, for example, that $F3$ corresponds to the principal direction of inertia of group 3 and to a direction of lesser inertia of group 2. Here we observe the exact structure used to construct the data.

Inertias of Individuals in the Mean Cloud

Whatever the axis, the four individuals have the same coordinate (in absolute value) and therefore the same contribution to inertia. Here we again see the symmetry of individuals which can be clearly seen in the data.

	F1	F2	F3
A	1.41	1.12	1.06
B	1.41	1.12	−1.06
C	−1.41	1.12	−1.06
D	−1.41	−1.12	1.06

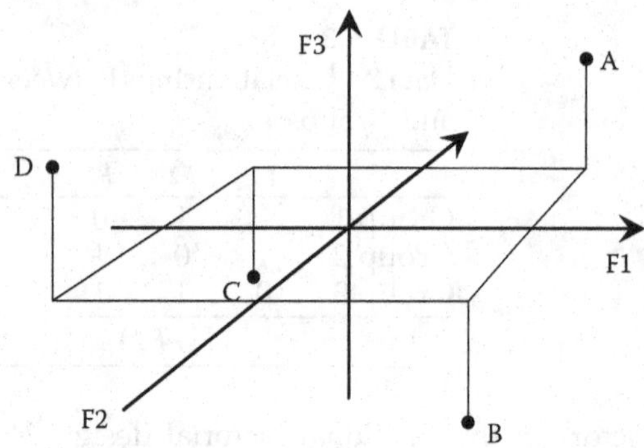

FIGURE 9.4
Data 2^{3-1}. MFA. Representation of the mean cloud. Roundings and exact values: $1.41 = \sqrt{2}$; $1.12 = \sqrt{5}/2$; $1.06 = \sqrt{10}/3$.

Superimposed Representation

This representation specifies the nature of the common factors (see Figure 9.5). The first factor opposes A and B on the one hand, with C and D, an opposition which exists in groups 1 and 2 and not in group 3. In this very particular example, we already saw this when we identified factor 1 with variable X.

In detail, this representation illustrates the distortions of the partial clouds in the superimposed representation. Indeed, although N_I is perfectly represented, clouds N_I^j are distorted according to the two aspects described in Section 5.5. Thus, in terms of the initial N_I^j, the representations of N_I^j in the MFA from the example present the following two characteristics:

1. N_I^j were subject to a homothety of ratio $1/\sqrt{\lambda_s}$ along each factor of cloud N_I. Thus, cloud N_I^1 initially possesses the same inertia in all directions but, when projected, is longer along $F2$. For this first group, the numerical calculation is simple as both of the nonzero eigenvalues for its separate PCA are equal. Between axes 1 and 2 of the MFA, the square of the ratio of the coordinates of partial individuals (coordinates with identical absolute values for all points and for axes 1 and 2) is the opposite of that of the eigenvalues of the MFA. Thus, numerically, as $\lambda_1 = 2$ and $\lambda_2 = 1.25$:

$$\left[\frac{F_2(i)}{F_1(i)}\right]^2 = \left[\frac{6}{\sqrt{5}}\frac{\sqrt{2}}{3}\right]^2 = \frac{\lambda_1}{\lambda_2} = \frac{2}{1.25}.$$

2. N_I^j were subject to a homothety of ratio $\sqrt{\lambda_s^j}$ along their own factors (λ_s^j being the sth eigenvalue in the separate analysis of N_I^j). Thus, for $j = 3$, the rectangle formed by N_I^j is more elongated in the projection than it was initially (the distance between the second and third

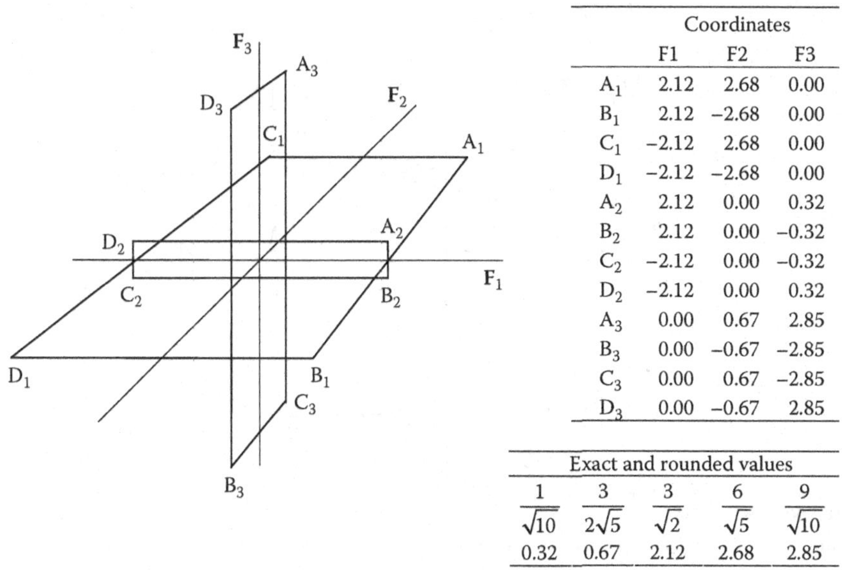

	Coordinates		
	F1	F2	F3
A_1	2.12	2.68	0.00
B_1	2.12	−2.68	0.00
C_1	−2.12	2.68	0.00
D_1	−2.12	−2.68	0.00
A_2	2.12	0.00	0.32
B_2	2.12	0.00	−0.32
C_2	−2.12	0.00	−0.32
D_2	−2.12	0.00	0.32
A_3	0.00	0.67	2.85
B_3	0.00	−0.67	−2.85
C_3	0.00	0.67	−2.85
D_3	0.00	−0.67	2.85

Exact and rounded values				
$\frac{1}{\sqrt{10}}$	$\frac{3}{2\sqrt{5}}$	$\frac{3}{\sqrt{2}}$	$\frac{6}{\sqrt{5}}$	$\frac{9}{\sqrt{10}}$
0.32	0.67	2.12	2.68	2.85

FIGURE 9.5

Data 2^{3-1}. MFA. Superimposed representation. The coordinates of N_I^j are multiplied by the number of active groups J (here 3) so that the overall cloud N_I might be at the centre of gravity of N_I^j.

eigenvalues of N_I is minimal and is of little importance here). For this third group, a simple calculation is possible as the factors of the MFA coincide, up to a rank, with those of its separate PCA: axis 2 (and 3, respectively) of the MFA coincides with axis 2 (and 1, respectively) of the separate PCA of group 3. Therefore, between axes 2 and 3, the square of the ratio of the coordinates of the individuals is simply expressed according to the eigenvalues of the MFA and those of the PCA of group 3. Thus, numerically, as $\lambda_1^3 = 1, \lambda_2^3 = 1/4, \lambda_2 = 1.25$ and $\lambda_3 = 10/9$:

$$\left[\frac{F_3(i)}{F_2(i)}\right]^2 = \left[\frac{9}{\sqrt{10}}\frac{2\sqrt{5}}{3}\right]^2 = \frac{\lambda_2}{\lambda_3}\frac{\lambda_1^3}{\lambda_2^3} = \left[\frac{5}{4}\frac{9}{10}\right]4.$$

9.3.3 Results of the GPA

Projected Inertias of the Mean Cloud

By construction, this cloud is contained within a plane. The percentages of inertia are 63.5% and 36.5%.

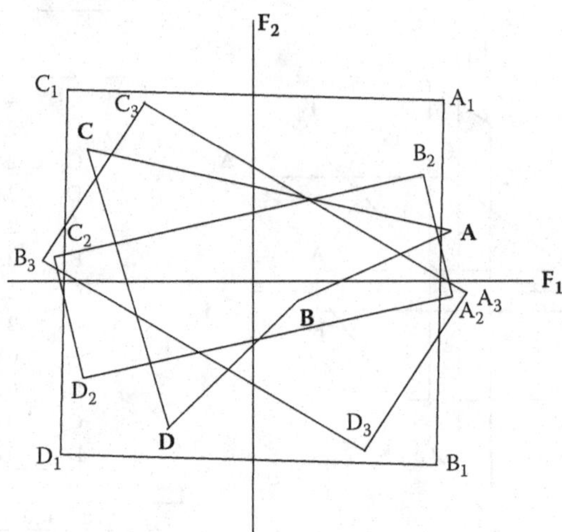

FIGURE 9.6
Data 2^{3-1}. GPA. Superimposed representation.

Representing Individuals

Each cloud N_I^j is represented perfectly (see Figure 9.6). Of course, the homologous points do not superimpose exactly as the data do not respect the Procrustes model exactly. Unlike N_I^j, there is no symmetry in the mean cloud: the individuals play very different roles. Of course this is not in accordance with the data.

That being said, we again see the key features of the first plane of the MFA, that is to say, a first axis which, overall, confronts $\{A, B\}$ with $\{C, D\}$ and a second which, overall, confronts $\{A, C\}$ with $\{B, D\}$.

Indicators of the Discrepancy to the Procrustes Model

The decomposition of total inertia into between-inertia and within-inertia is examined overall and then decomposed in different ways (see Table 9.4). Total inertia is fixed at 100.

Overall Decomposition. The between-inertia (58.3) is:

- Substantially lower than 100: The Procrustes model is far from verified.
- Substantially higher than 100/3: The three clouds have common structural elements.

Decomposition by Dimension. The between-inertia is stronger for axis 1 (37.0 > 21.3). The contrast between $\{A, B\}$ and $\{C, D\}$ has a strong inertia in one cloud (the 2) which is not the case of the contrast between $\{A, C\}$ with $\{B, D\}$. But

TABLE 9.4
Data 2^{3-1}. GPA. Decompositions of Total Inertia

Decomposition	Between-Iner.	Within-Iner.	Total
Overall	58.3	41.7	100
By dimension			
1	37.0 (58%)	26.2	63.3 (100%)
2	21.3 (56%)	15.5	37.7 (100%)
By group			
1		13.3 (29%)	45.9 (100%)
2		9.6 (38%)	25.5 (100%)
3		18.8 (65%)	28.7 (100%)
By individual			
A	20.6	4.4	25
B	1.4	23.6	25
C	21.4	3.6	25
D	15.0	10.1	25

the percentage of between-inertia is almost the same for both dimensions. The consensus degree is similar between the two axes.

Decomposition by Group. The mean cloud (which is interpreted as the common structure) is most similar to groups 1 and 2 and much less so to group 3 (stronger within-inertia for this group: 18.8). This decomposition suggests that groups 1 and 2 are more alike than they are like group 3. This can be seen in the data (see Figure 9.3) and in MFA (see Table 9.3: the common direction between N_I^1 and N_I^2 is a direction of maximum inertia).

Decomposition by Individual. The different representations of individuals A and C are grouped closely around their centre of gravity (low within-inertia: 4.4 and 3.6), unlike those of individual B which are far apart. This suggests that individual B plays a particular role in the data, which contradicts the symmetries in the data. In fact, if, in the configuration of the GPA, we switch individuals *A* and *D* and individuals *B* and *C*, we obtain a solution which is just as good as the previous one but which this time suggests that it is individual *C* which is very particular. This instability is not satisfactory.

Conclusion

In this example of very low dimension, the constrained nature of the model heavily influences the GPA of the data which does not verify the Procrustes model. The specificities of the groups make it difficult to identify the common structures clearly.

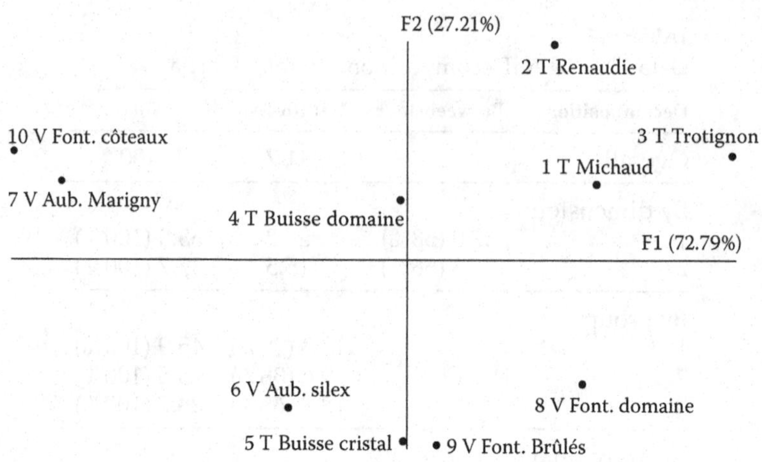

FIGURE 9.7
Wines. GPA. Representation of the mean cloud.

9.4 Application to the Ten Touraine Wines

These data have already been analysed in Chapter 7. In these napping® data, 11 tasters each provided a planar configuration of 10 wines.

We are here dealing with a GPA in its true context. As the initial configurations are planar, they can be represented perfectly on a plane. Otherwise, when they have more than two dimensions, the advantage of the GPA constraint of nondistortion of the configuration in looking for a superimposed representation becomes unclear, as it is not possible to represent the configurations without distorting them (due to their projection on the axes of N_I).

We do not reproduce the superimposition of the 10 configurations as it is extremely overcrowded and only really of interest to users who know the tasters well. The mean configuration generated by the GPA (see Figure 9.7) is very similar to that generated by the MFA (see Figure 7.7). The RV between these two planar configurations is .906. More precisely (see Table 9.5), these two representations have almost the same first axis ($r = .9917$), which is not the case for the second axis ($r = .8734$). As a result, the first bisector in GPA does not clearly separate the two vines (as it does in MFA).

The second axis of the GPA isolates wines 5, 6, 8 and 9 from the others by bringing them closer to one another. This does not seem to make sense. In MFA, wines 8 and 9 mainly contribute to the second axis (60%) and wines 5 and 6 mainly contribute to the third axis (73%). Unlike the MFA, the GPA (in this application) only has two dimensions. This is why this second axis of the GPA is difficult to interpret. The impossible compromise (roughly between axes 2 and 3 of the MFA), that is the second axis of the GPA, also induces a difference in inertia, with the first axis, which is very high compared to that of the MFA ($72.79 - 27.21 > 36.39 - 26.68$).

TABLE 9.5
Wines. Correlation Coefficients[a]

	MFA1	MFA2	MFA3	GPA1	GPA2
MFA1	39.39%				
MFA2	0	26.68%			
MFA3	0	0	11.45%		
GPA1	−0.9917	0.0147	0.0716	73.15%	
GPA2	−0.0581	−0.8734	−0.4381	0	26.85%

[a] Between the factors of the MFA and those of the GPA. The percentages of inertia associated with the factors can be seen on the diagonal.

In this application on real data, we can see the drawback of GPA identified in the 2^{3-1} data. In fact, the advantage of GPA resides in the superimposed representation. The mean cloud is merely an intermediary calculation and its use as a mean cloud is limited.

On the other hand, the superimposed representation of the MFA distorts the partial clouds (see Section 5.5). This is problematic in an application such as napping® in which each taster expects to find his or her exact nappe (on the factorial plane). This is the origin of Procrustes MFA (PMFA) which, from the mean configuration of the MFA, conducts a Procrustes rotation of each partial configuration on this mean configuration.

For example, Figure 9.8 shows the resulting graph for the PMFA of nappes 8 and 9. This is therefore a Procrustes rotation applied to the configurations of Figure 7.6 in order to make them coincide as closely as possible with the configuration of the wines from Figure 7.7.

It is not surprising to see that the largest dimension of nappe 9 (and 8, respectively) coincides (more or less) with the first (and, respectively, second) axis: this supports the representation of the tasters generated by the MFA (see Figure 7.7, right). Taster 9 (and 8, respectively) attributes great importance (in terms of the Indscal model) to the dimension illustrated by the first (and, respectively, second) axis of the MFA.

This clearly illustrates an advantage which has been mentioned many times in this work, that is to say, the fact that each perspective of the MFA (here, that of the superimposed representation, the GPA), is related to other perspectives (that of Indscal in the above comment).

9.5 Conclusion

MFA and GPA are fundamentally different methods with different objectives. In particular, GPA constructs a representation of the data within a constrained framework (each dimension is common to all of the groups). GPA must be reserved for specific applications.

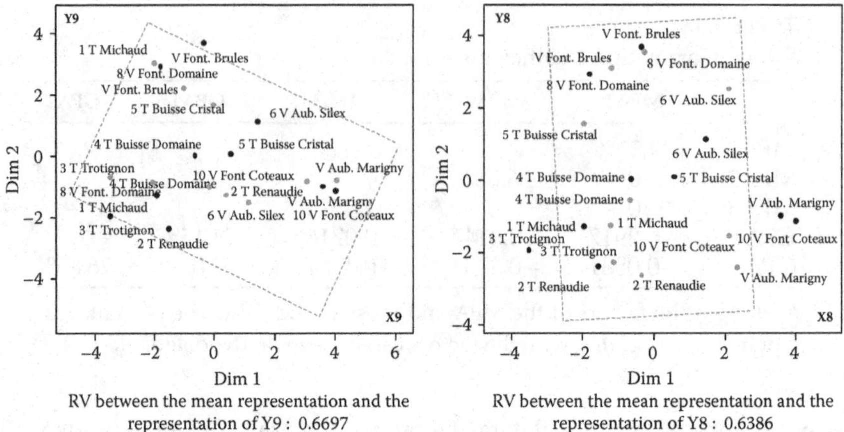

FIGURE 9.8
Wines. PMFA. Representation of nappes 8 and 9 on the mean configuration of the MFA.

Concretely, GPA generates an exact superposed representation of N_I^j. The mean cloud is merely an intermediary used to obtain the superimposed representation. Its representation can inappropriately express a common structure when accompanied by specific structures. We do not recommend using GPA to obtain a mean configuration. MFA, on the other hand, focusses on representing the mean cloud and generates a representation which fully benefits by the dual nature of factorial analysis. However, the representation of N_I^j is distorted.

In practice, we are often looking for dimensions common to groups of variables on data which, aside from a few common dimensions, have many dimensions which are both non-common and have low inertia. In these situations, GPA's limitations are not a problem as the non-common dimensions which it superimposes concern the dimensions of low inertia which are not examined. This explains the convergence of the results which can be observed in practice on the first plane.

9.6 GPA in FactoMineR

Here we use the 2^{3-1} data. There are no specific instructions for implementing GPA via **R Commander** as it should pose no problem for users. The default values of the algorithm's parameters are (almost) always suitable in practice.

Below are some of the command lines used for this chapter.

```
# Reading and verifying data
> D2=read.table("DON2(3-1).csv",header=T,sep=";",row.names=1)
> D2
```

```
  X1 Y1 X2 Z2 Y3 Z3
A  6  6  6 -2  3 -6
B  6 -6  6  2 -3  6
C -6  6 -6  2  3  6
D -6 -6 -6 -2 -3 -6
```

As in MFA, the variables of a given group must be consecutive in the file and the groups are defined by their number of variables. Only quantitative variables can be used. The data are always centred. Reduction is optional but is applied to all groups in the same way. In the example, the variables must not be reduced:

```
> res=GPA(D2,group=c(2,2,2),scale=F)
```

By default, this instruction displays the mean and partial points in a *star representation*. It includes the `plot.GPA` function, with all options selected by default:

```
> plot.GPA(res)
```

The colours of the partial points are generally chosen according to the groups:

```
> plot.GPA(res,hab="group")
```

The graph thus generated was used as a basis for creating Figure 9.6 along with graphics software. The `res` list contains the different decompositions of the inertia. The raw values for Table 9.4 are obtained using the following code.

```
# Initialisation of table 9_4
> tab9_4=matrix(nrow=10,ncol=3)
# Choice of names for rows and columns
> colnames(tab10_4)=c("Between In.","Within In.","Total In.")
> row.names(tab9_4)=c("Global","Dim1","Dim2","G1","G2","G3",
+ "A","B","C","D")
# All the decompositions of inertia are in the
# tables in the list res$PANOVA
# By dimension
> tab9_4[1:3,]=res$PANOVA$dimension[c(3,1,2),]
# By group
> tab9_4[4:6,]=res$PANOVA$config[1:3,]
# By individual
> tab9_4[7:10,]=res$PANOVA$objet[1:4,]
# Edition
> round(tab9_4,1)
```

Procrustes MFA (PMFA; Figure 9.8)
This method is limited to sensory data collected using napping®. It is not available in **FactoMineR** but in the R package **SensoMineR**. In this package,

PMFA is available using the `pmfa` function. This function can be applied to a data table with the same format as that of MFA with, as a constraint, each group being made up of two (consecutive) quantitative variables. In the case of the Touraine wines (see Sections 9.4 and 7.6.4), this means duplicating the single dimension of taster 10. These constraints make it possible to implement the function with all the default options:

```
# Data importation (napping with white wines from Loire Valley)
> napping=read.table("napping.csv",header=TRUE,sep=";",
+ row.names=1)
# PMFA function is in SensoMineR
> library(SensoMineR)
# Each nappe must have 2 dimensions;
# hence the duplication of column 19 (nappe 10)
> library(SensoMineR)
> res=pmfa(napping[,c(1:19,19:21)])
```

Below is the code used to process the GPA of these data (see Figure 9.7 and Table 9.5).

```
# GPA
> library(FactoMineR)
> res=GPA(napping[,c(1:21)],group=c(rep(2,9),1,2),scale=FALSE)
# Figure 9.7
> plot(res,partial="none")
# The inertias by dimension are in res$PANOVA$dimension
# To obtain the percent of inertia of the mean cloud
> res$PANOVA$dimension[1:2,1]/res$PANOVA$dimension[3,1]*100

# Table 9.5
# Concatenation of the 3 first factors of MFA and of the 2 of GPA
> resmfa=MFA(napping[,c(1:19,19:21)],group=rep(2,11),
+ type=rep("c",11),graph=F)
> comp=cbind(resmfa$ind$coord[,1:3],res$consensus)
> colnames(comp)=c("MFA1","MFA2","MFA3","GPA1","GPA2")
> Tab9_5=cor(comp)

# Percentages in the diagonal
> percentmfa=resmfa$eig[1:3,2]
> percentgpa=res$PANOVA$dimension[1:2,1]/res$PANOVA$
+ dimension[3,1]*100
> diag(Tab9_5)=c(percentmfa,percentgpa)
> round(Tab9_5,4)
```

10

Hierarchical Multiple Factor Analysis

Hierarchical multiple factor analysis (HMFA) is the most direct extension of multiple factor analysis (MFA): it is used with tables in which the variables are structured according to a hierarchy. In practice, this means a sequence of nested partitions. At first sight, one might be tempted to think that this is merely a curiosity, but HMFA has considerable application potential due to the increasingly complex nature of the data users wish to analyse simultaneously.

10.1 Data, Examples

We often want to analyse individuals $\times$ variables tables in which the structure of the variables is more complex than a simple partition. To illustrate this point, let us look back at the orange juices.

In fact, in these data, in addition to the chemical and sensory measurements, we also have overall evaluation scores (known as *hedonic assessments* in the field of sensory analysis) for each of the six juices, attributed by 96 consumers. We want these data to indicate which of these juices the consumers preferred. Using the terminology utilized in the presentation of principal component analysis (PCA), we refer to the *hedonic profile* of a product as all of the assessment scores it obtained. This table should therefore make it possible to identify juices with similar hedonic profiles (when someone likes one juice, she likes the other) or opposite profiles (when someone likes one juice, she does not like the other). These scores can be presented in a table with the six juices in the rows and the consumers in the columns, with, at the intersection of row i and column k, the score given by consumer k for product i. Presented in this way, that is to say, with the products associated with rows of a table, these data can be juxtaposed with the chemical (8 variables) and sensory (7 variables) data. In the end, we obtain a table with $8 + 7 + 96 = 111$ columns.

We might consider conducting an MFA of this table, with the three groups of variables introduced as active but this methodology would make each of the three groups play the same role, which would not meet the users' needs. Users primarily want to link the hedonic assessments to the characteristics of the products, these characteristics being chemical measurements or sensory descriptors. This perspective suggests the hierarchical structure on the data

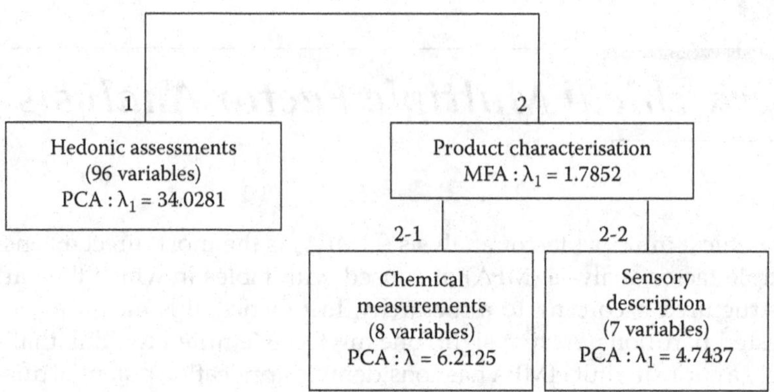

FIGURE 10.1
Data structure. The nodes of the tree are numbered like the chapters of a book (see text). For each, we specify the method (PCA or MFA) corresponding to the analysis of the variables brought together by the node, as well as its first eigenvalue.

illustrated in Figure 10.1. The aim of this chapter is to show what accounting for a hierarchical structure on the variables means.

The Bac data, used to illustrate the PCA in Chapter 1, provides a second example of a hierarchy of variables. In fact, in these data, along with each student's Baccalaureate results, we also have their term grades (there are three terms) for each of the five subjects. The complete data table therefore has 5 (subjects) $\times$ {1 (Bac) + 3 (terms)} = 20 columns. It is possible to conduct many interesting analyses on these data. An overall perspective is to account for the hierarchy of variables which first separates the five Bac grades from the fifteen term grades and, within these term grades, the three blocks of five term grades.

These two examples illustrate a common source of hierarchies defined from variables: collecting different kinds of data for one set of statistical individuals. Surveys are another such source: questionnaires are often organised into themes and subthemes; it is interesting to take this structure into account in the analysis.

10.2 Hierarchy and Partitions

Figure 10.2 contains two representations of one hierarchy defined on a set of seven variables $\{X_1, \ldots, X_7\}$. In these graphs, a node is represented by a horizontal segment. The highest segment represents the *root node.*

In our context, it is helpful to consider a hierarchy as a sequence of nested partitions, remembering that, in practice, only some of these partitions will be

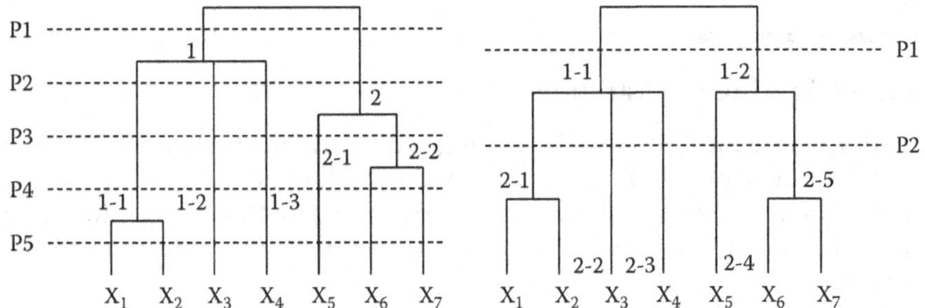

FIGURE 10.2
Two representations of one hierarchy of seven variables.

of interest. This is illustrated in Figure 10.2, which identifies two sequences of partitions from one single hierarchy. In our applications, the hierarchies are not indexed: as Figure 10.2 shows, graphically introducing indexes to each node makes it possible to visualise a specific sequence of partitions.

These are not merely formal considerations. In practice, that is to say, in the function which conducts an HMFA in the **FactoMineR** package, the hierarchy is introduced using a sequence of nested partitions. The results concerning the nodes are given according to this sequence. In Figure 10.2, the sequence on the left makes it possible, via partition $P2$, to determine the components of node 1 (X_1, X_2, X_3, X_4) without identifying those of node 2 (X_5, X_6, X_7), which is not the case for the sequence on the right. This aspect is particularly important in the graphs representing partial individuals, which quickly become overcrowded (see below).

We show that, in the calculation procedure as well as that used to determine the hierarchy, we proceed from the bottom upwards, from the most to the least detailed partition. However, when analysing the results, we proceed in the opposite sense, from top to bottom. It is therefore useful to number the partitions from the top down (see Figure 10.2).

When interpreting the analysis, we constantly refer back to the groups of variables defined in the tree, that is to say, the nodes of the tree. It is therefore important to label them. There are two ways of labelling the data, bearing in mind that the root node, which includes all of the variables, does not need to be labelled.

One system (see Figure 10.2, right) is based on the partitions. A node receives a label concatenating the rank of the partition which defines it, followed by an (arbitrary) ordinal number in the partition. Therefore, in this system, 2-1 is the first node of the partition at level 2.

Another system (see Figure 10.2, left, and Figure 10.1) expresses the path between the root node and another given node (in much the same way as the sections of a book might be numbered, like in this one). Therefore, in this system, 2-1 is the first node (directly) dependent on node 2 (which itself is directly attached to the root node).

10.3 Weighting the Variables

In MFA, taking into account partitioning of the variables first means balancing the role of the groups in an overall analysis. This idea is transposed in the case of a hierarchy of the variables: for each node, taking this hierarchy into account means balancing the role of the groups of variables which descend directly from that node.

Therefore, in the Orange Juice data, the *chemical* and *sensory* groups first need to be balanced. This balance must be conducted in the sense of MFA, that is to say, by fixing the maximal inertia at 1. Firstly, we endow these variables with their weights in the MFA of these two groups (that is to say, the inverse of the first eigenvalue of the PCA of their group; λ_1^n is the first eigenvalue of the analysis of node n) thus, for a sensory variable (belonging to node 2-2; see Figure 10.1) $1/\lambda_1^{2\text{-}2} = 1/4.7437 = .2108$ and, for a chemical variable (belonging to node 2-1), $1/\lambda_1^{2\text{-}1} = 1/6.2125 = .1610$.

In addition, the groups' *characterisation* and *hedonic assessments* need to be balanced. In order to do this, the hedonic assessments (node 1) will be weighted by $1/\lambda_1^1 = 1/34.0281 = 0.02939$, and the characterisation variables will be overweighted by the inverse of the first eigenvalue of their MFA, $1/\lambda_1^2 = 1/1.7852 = .5602$. They are said to be overweighted as these variables are already weighted in the MFA associated with node 2. Finally, the weights associated with the characterisation variables are:

- Chemical variables: $\left(1/\lambda_1^{2\text{-}1}\right)\left(1/\lambda_1^2\right) = .0902$
- Sensory variables: $\left(1/\lambda_1^{2\text{-}2}\right)\left(1/\lambda_1^2\right) = .1181$

Using these weights in the overall analysis:

- The groups' *characterisation* and *hedonic assessment* are balanced in terms of the MFA (in the full sense of the word: the maximum axial inertia of each group is fixed at 1).
- The groups' *chemical* and *sensory* are balanced in terms of the MFA (in a limited sense: the maximum axial inertias are all equal, but are not 1, more precisely $1/\lambda_1^2$).

More generally, let us consider a variable k, with an initial weight of p_k^0 (generally $p_k^0 = 1$) and the sequence of nodes here denoted $\{0, 1, \ldots, n, \ldots, N\}$ (node 0 corresponds to the variable considered alone) linking it to the root node.

Let λ_1^n be the first eigenvalue of the factorial analysis associated with node n. In the case of quantitative variables, this analysis is a PCA if $n = 1$, an MFA if $n = 2$ and an HMFA if $n > 2$. The weight of variable k in the analysis associated with any node n_0 is

$$p_k^{n_0} = p_k^0 \prod_{n=1}^{n=n_0} \left(\frac{1}{\lambda_1^n}\right).$$

TABLE 10.1
Orange Juice. HMFA. Decomposition of Inertia by Axis and by Node

Node		F1	F2	F3	F4	F5
	HMFA	1.934 (44.25%)	0.775 (17.73%)	0.648 (14.82%)	0.512 (11.71%)	0.501 (11.46%)
1	Hedonic assess.	0.957	0.577	0.489	0.412	0.386
2	Characterisation	0.976	0.198	0.159	0.100	0.115
2-1	Chemical	0.449	0.060	0.134	0.068	0.010
2-2	Sensory	0.527	0.138	0.026	0.031	0.104

We simply denote p_k the weight of variable k in the final analysis.

Algorithm. An HMFA can be conducted using an MFA program (and therefore PCA). The tree is scanned from the bottom up. For each node, an MFA is conducted on the variables it brings together and these variables are replaced by the (unstandardised) factors of this MFA. The MFA associated with the root node provides the factors of the HMFA. The HMFA in **FactoMineR** proceeds in this way.

Table 10.1 summarises the decompositions of inertia in the HMFA of the Orange Juice data. This shows how the HMFA balances the influence of the hedonic assessments and characterisation on the one hand (rows 3 and 4) and of chemical and sensory (last two rows). This balance is excellent for the first axis ($.957 \approx .976$ and $.449 \approx .527$). For the following axes, the predominance of hedonic assessments derives directly from a more homogeneous distribution of inertia in this group.

Remark
In the Orange Juice data, the variables are quantitative. In the presence of qualitative variables, in the above reasoning the PCA should be replaced by multiple correspondence analysis (MCA). When the less-detailed partition includes mixed groups, PCA is then replaced by factorial analysis of mixed data (FAMD).

10.4 Representing Partial Individuals

10.4.1 Method

In MFA the partial cloud (of individuals) associated with group j (denoted N_I^j) is obtained by projecting the mean cloud (denoted N_I) on the subspace (of $\mathbb{R}^K$) generated by the variables of group j alone. The same principle is used in HMFA with the groups' roles played by the nodes. Thus, the partial cloud of node n (denoted N_I^n, it is made up of partial individuals denoted i^n) is the projection of N_I on the subspace generated by the variables which depend on node n alone.

The representation of cloud N_I^n is obtained in the same way as for MFA, that is to say, by projection on the principal axes of N_I. This representation benefits from the same properties as in MFA, particularly the partial transition relation which expresses the coordinate of point i^n according to the coordinates of the variables depending on node n.

The notations from MFA (see Section 5.2) are adapted to HMFA.

K_n: Variables dependent on node n.

F_s^n: Factor of rank s partial to node n. This factor contains the coordinates of the partial individuals i^n on the axis of rank s (of mean cloud N_I). Its ith term is denoted $F_s^n(i)$ or $F_s(i^n)$ interchangeably.

M: (Diagonal) Metric in $\mathbb{R}^K$ for the overall analysis. Unlike in MFA, the weights (p_k) of the variables from a given group defined by a given node are not generally identical.

With these notations, the partial transition relation of the MFA (see Section 5.2) is transposed directly:

$$F_s^n = F_s\left(i^n\right) = \frac{1}{\sqrt{\lambda_s}} \sum_{k \in K_n} x_{ik} p_k G_s(k).$$

The interpretation rule resulting from this relation is the same as for MFA: on the graphs, partial individual i^n is on the side of the variables (of node n) for which it has a high value and opposite the variables (of node n) for which it has a low value. The only difference between the two relations is the presence of weights of variables (p_k) which are not generally constant within a node.

On the MFA graphs, the mean point i is at the barycentre of the partial individuals (hence the name *mean point*). To obtain this property, cloud N_I^j is dilated with the J coefficient (see Section 5.2).

This property can be transposed to HMFA as follows: each partial point i^n must be at the barycentre of the partial points which are directly dependent on node n. In the example of the six orange juices:

- Mean point i must be at the barycentre of its partial points *characterisation* and *hedonic assessments*.
- Partial point i *characterisation* must be at the barycentre of its partial points *chemical* and *sensory*.

As in MFA, this property is obtained by dilating clouds N_I^n. For the J nodes dependent on the root node, dilation is conducted in the same way as in MFA, using the J coefficient. For the other nodes, it is important to account for the dilations associated with the nodes that link them back to the root node.

Thus, in the Orange Juice example, the partial points *chemical* and *sensory* must be dilated with coefficient 2 so that the partial point *characterisation*

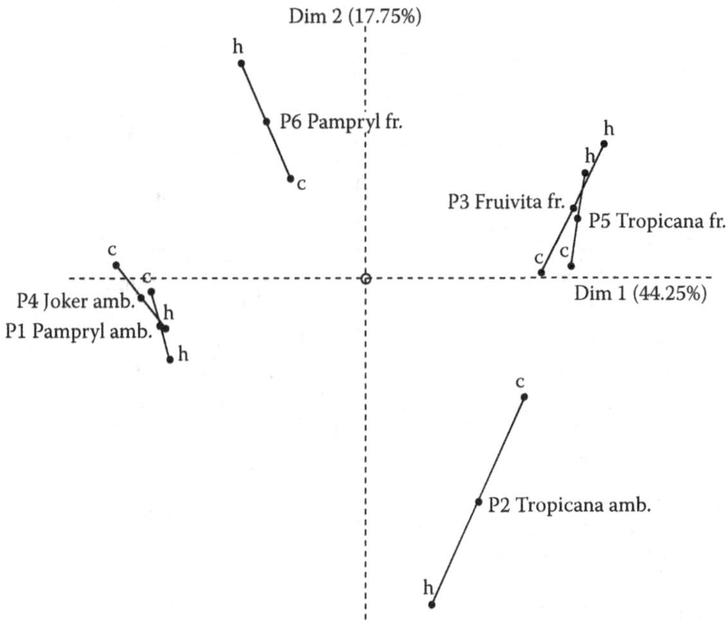

FIGURE 10.3
Orange Juice. HMFA. Representation of partial points corresponding to the *characterisation* (c) and *hedonic assessment* (h) nodes on the first plane.

might be at their barycentre. But this partial point *characterisation* was itself dilated, with coefficient 2, so that mean point i might be at the barycentre of its partial points *characterisation* and *hedonic assessments*. Finally, the partial clouds *chemical* and *sensory* will be dilated with coefficient $4 = 2 \times 2$.

More generally, denoting J_n the number of direct descendants of node n, that is to say, the number of elements which are directly related to it, and A_n all of the ascendants of node n (including the root node), the dilation coefficient of N_I^n is worth

$$\prod_{n \in A_n} J_n.$$

10.4.2 Application to the Six Orange Juices

Figure 10.3 represents the partial points for the *hedonic assessment* and *characterisation* nodes only; Figure 10.4 identifies the *chemical* and *sensory* aspects.

First of all, the representation of mean individuals resulting from the HMFA is similar to that generated by the MFA of the characterisation data alone (see Figure 4.4). From the sensory and chemical perspective, it therefore supports the same interpretation (see representation of variables in Figure 4.4). The new aspect here stems from the hedonic assessments: the opposition 'hard

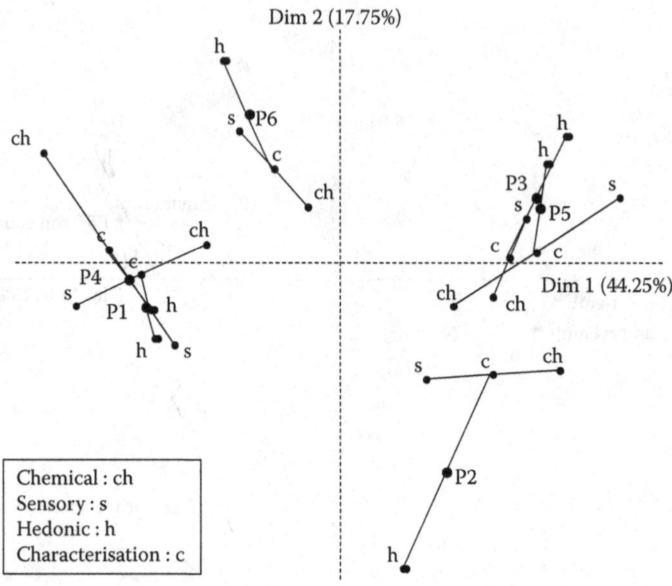

FIGURE 10.4
Orange Juice. HMFA. Representation of all the partial points.

juices' ↔ 'soft juices' is correlated with many hedonic assessments. We can go further still: the projection of the axes of the separate analyses of the nodes (see Figure 10.5) shows that this first factor of the HMFA is very similar to the first factor of the PCA of the hedonic assessments alone (the correlation coefficient between these two factors is .957). This convergence between principal dimensions of characterisation and hedonic assessments is absolutely remarkable.

The superimposed representation shows that, from the perspective of the first axis (opposition 'soft juices' ↔ 'hard juices'):

- Juices 2, 3 and 5 are the same from the *characterisation* perspective; this can be seen in the MFA of the characterisation data alone (see Figure 4.4).
- Juices 3 and 5 are much more typical than 2 from the *hedonic assessment* perspective; (compared to 3 and 5) the latter was much less appreciated by those preferring soft juices and/or much less badly scored than those who like hard juices (this can be clearly seen on the PCA of the hedonic assessments alone, which is not reproduced here).

This similarity of the overall characterisations of juices 2, 3 and 5 hides disparities when the chemical and sensory aspects are detailed (see Figure 10.4). Juice 2 is more typical of a soft juice from a chemical perspective than from a sensory point of view. The opposite is true for juice 3. This observation

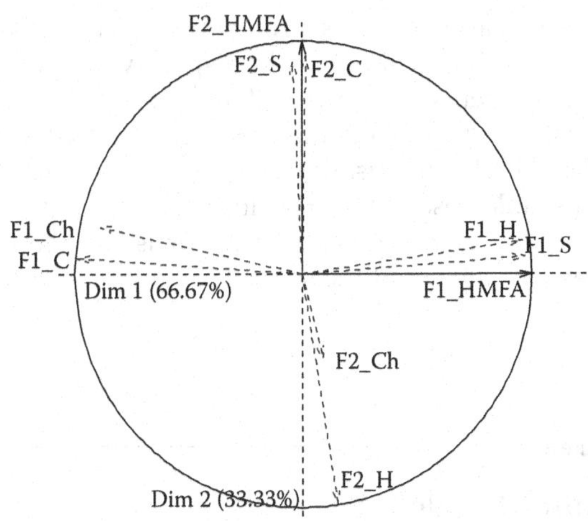

FIGURE 10.5
Orange Juice. HMFA. Representation of the axes of the separate analyses of the nodes. H: Hedonic assessments; C: Characterisation; Ch: Chemical; S: Sensory.

corresponds well to the plane of the MFA on the characterisation data alone (see Figure 4.4).

This illustration shows that the rules for interpreting the representations of partial points in HMFA are identical to those of MFA.

10.5 Canonical Correlation Coefficients

As in MFA, it is useful to measure the similarity between the representation of the mean cloud and that of each partial cloud. We can thus calculate the correlation coefficient between F_s and F_s^n for all n and all the first axes. We obtain Table 10.2, analogous to Table 6.1 in MFA.

The first two rows remind the user of the problems related to the dimensions of these data, that is to say, the small number of individuals compared

TABLE 10.2
Orange Juice. HMFA. Canonical Correlation Coefficients

	F1	*F2*	*F3*	*F4*	*F5*
Hedonic assessments	0.989	0.994	0.967	0.997	0.994
Characterisation	0.990	0.952	0.776	0.958	0.938
Chemical	0.902	0.492	0.623	0.550	0.325
Sensory	0.972	0.913	0.400	0.281	0.871

with the number of variables. This problem is particularly notable for the *hedonic assessments* node (96 variables in a space with five dimensions) for which the five coefficients are greater than .966. In $\mathbb{R}^I$, every direction is a linear combination of the hedonic assessments. This is also true, to a lesser extent, for the characterisation node (15 variables). Finally, the last two rows of the table resemble those in Table 6.1 of the MFA for characterisation alone. In summary, the first factor is common to the hedonic, chemical and sensory data, and the second to only the hedonic and sensory data.

10.6 Representing the Nodes

In MFA, the representation of groups of variables (known as the *relationship square*; see Section 7.3) presents an interesting property: the coordinate of an active group j along axis s is interpreted both as the relationship measurement Lg between group j and the sth factor, and as the contribution of group j to the construction of the axis of rank s.

In HMFA this property only exists for the nodes linked directly to the root node (HMFA relies on an MFA of these nodes). For the others, these two notions do not coincide. In this type of graph, it is natural to favour the notion of relationship to enable the representation of supplementary elements. With this choice, a set of variables which appear twice in the hierarchy lies in one place.

In the example of the orange juices, this representation (see Figure 10.6) shows that:

- The first axis corresponds to a direction of high inertia for each of the nodes; this is not new: the strong correlation between this axis and the first component of the separate analyses of the nodes has already been mentioned.
- The second axis is more closely related to hedonic assessments than characterisation; this is due to the higher dimensionality of the hedonic assessments (see Table 10.1). As characterisation (and particularly its two aspects) has a predominant dimension linked to the first axis, it cannot be closely related to the other dimensions in terms of Lg.
- The relationship with the first axis is stronger for the sensory group than for the chemical group. In the MFA on the characterisation data alone (see Figure 7.3 and Table 4.5), this relationship was at the same level. This result is probably the consequence of the influence of the hedonic data, more directly linked to the sensory data than to the chemical data: 0.8197 > 0.6576 (see Table 10.3).

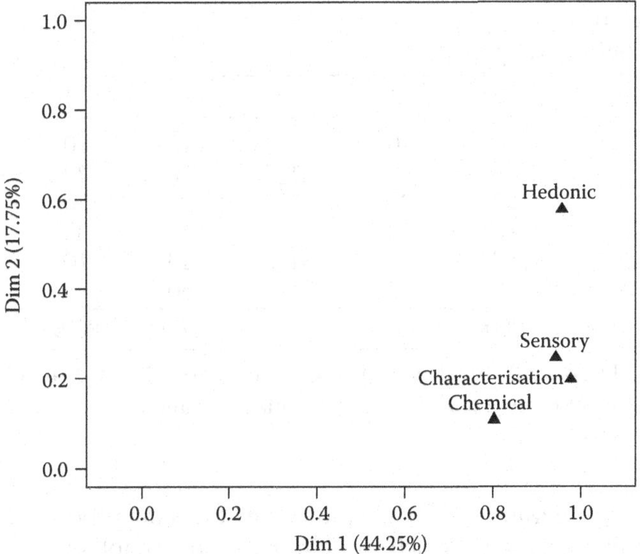

FIGURE 10.6
Orange Juice. HMFA. Representing the nodes (relationship square).

10.7 Application to Mixed Data: Sorted Napping®

10.7.1 Data and Methodology

In sensory evaluation, classic data collection involves asking judges (that is to say, tasters) to evaluate a set of products using a set of descriptors (variables). In this way products are attributed a sensory characterisation: a given product is perceived as sour, mildly bitter and so on. This classic approach does not take into account the importance of these descriptors in constructing the tasters' overall assessments.

To reveal the importance of each criterion, holistic methods can be used in which the judges directly evaluate the similarities between products, each according to his or her own criteria. There are three possible methods.

TABLE 10.3
Orange Juice. HMFA. RV Coefficients Between the Nodes

RV	MFA ch. + s.	Hedonic Assmt.	Chemical
MFA ch. + sensory	1		
Hedonic Assmt.	0.8245	1	
Chemical	0.8938	0.6576	
Sensory	0.9011	0.8197	0.6109

TABLE 10.4
Two Sorted Nappes[a]

Iden.	X1	Y1	S1	X2	Y2	S2
a	10	10	S1_1	10	10	S2_1
b	20	10	S1_1	10	20	S2_1
c	20	30	S1_2	20	30	S2_1
d	40	30	S1_2	40	30	S2_2
e	40	10	S1_3	40	10	S2_2
f	50	10	S1_3	50	10	S2_2
Variance	200	88.9	-	200	80.6	-

[a] X1: Horizontal dimension of judge 1 (in cm). S2_1: Group 1 of the sorting of taster 2 (containing individuals a, b and c).

Free sorting (= categorisation). Each judge partitions the products so that each class contains products which are similar to one another (in the judge's opinion) and between the classes where the products differ. In this data-collection method, each judge's data are a qualitative variable (defined on the products).

Napping®. Each judge provides a planar representation of products in such a way that two products are closer together (or farther apart, respectively), the more similar they are (or different, respectively). Originally, each judge physically laid out the products on a large sheet of paper 40 cm × 60 cm (a tablecloth, or *nappe* in French, hence the term *napping®*; in practice, the term *nappe* designates a configuration of products provided by a judge). Nowadays, this operation is often conducted on a screen: using specially designed software, the judge positions icons representing the products. In this data collection method, each judge's data are a pair of quantitative variables (the products' coordinates).

For the distances to be respected, the variables are not standardised when the data are analysed.

Sorted napping® combines the two previous approaches. After having positioned the products on the nappe, the judge groups together the products which he believes to be particularly similar. In this method, the data from each judge include three variables (the set of these three variables is a *sorted nappe*): a pair of quantitative variables (which make up what we refer to as the nappe) and a qualitative variable (sorting).

To illustrate the factorial analysis of such data, we use a small example made up of two sorted nappes (see Table 10.4 and Figure 10.7).

The aim of factorial analysis with this type of data is to obtain graphical representations of the objects at hand: the individuals, quantitative variables,

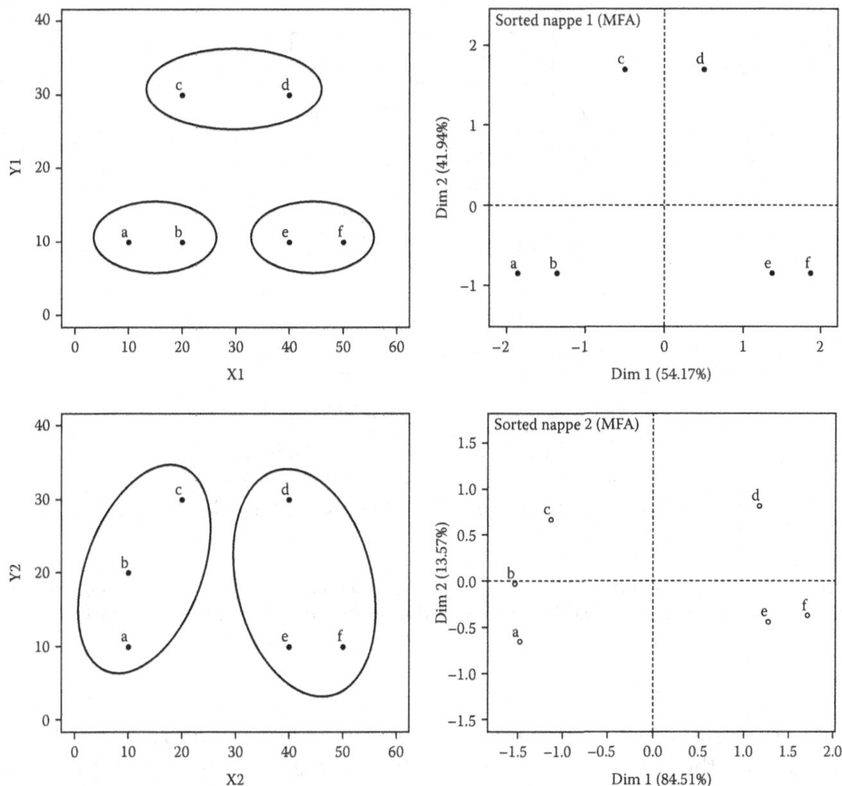

FIGURE 10.7
Two sorted nappes. 'Raw' nappes (left); nappes seen by their MFA (Euclidian representations of the sorted nappes, right).

categories of qualitative variables and sorted nappes. To conduct this analysis, it is important to balance:

- The two sorted nappes one against the other
- The nappe and the sorting within each sorted nappe

Balancing in these ways corresponds to the hierarchical structure of Figure 10.8. HMFA is the factorial analysis which takes this structure into account.

10.7.2 Intermediary Analysis: MFA on a Sorted Nappe

Before describing the results of this analysis, it is helpful to examine how the nappe and the sorting are taken into account simultaneously to obtain a sorted nappe. The HMFA proceeds in exactly this way as it is the same

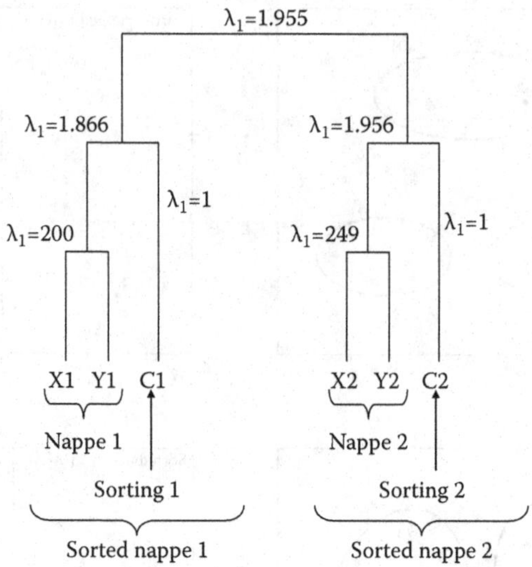

FIGURE 10.8
Two sorted nappes. Hierarchy of the six variables. Each node is attributed the first eigenvalue of the factorial analysis (PCA, MCA, MFA or HMFA depending on the cases) of the variables that it brings together.

as conducting a factorial analysis (PCA, MCA, FAMD, MFA or HMFA depending on the cases) for each of the nodes in the hierarchy, from the bottom of the tree upwards. The type of node considered here brings together the data of just one judge, that is to say, a group of two quantitative variables and one qualitative variable. This is therefore a good example in order to illustrate:

- In terms of statistical methodology: An MFA of quantitative and qualitative variables
- In terms of sensory methodology: The influence, in the analysis, of superimposing a sorting on a nappe.

Figure 10.7 (right) contains the first factorial plane of the MFA of each sorted nappe: this first plane is the Euclidean represention of the sorted nappe. Overall, in comparison with the nappe, the individuals of a given group (of sorting) are closer together on the sorted nappe (the term *sorted nappe* also designates Figures 10.7 left, which illustrate the nappe itself and the sorting, and Figures 10.7 right, Euclidean representations which account for both types of data).

The result is a sorted nappe 1 with two directions of comparable inertia (the horizontal and vertical dimensions of the nappe both correspond to an opposition between the classes of the sorting) and a sorted nappe 2 with a predominant direction of inertia (only the horizontal dimension corresponds

TABLE 10.5
Two Sorted Nappes. HMFA. Decompositions of Inertia[a]

	F1	F2	F3	F4	F5	Sum
HMFA eigenvalue	1.955	0.919	0.107	0.036	0.012	3.029
Sorted nappe 1	0.977	0.772	0.079	0.016	0.002	1.846
Sorted nappe 2	0.978	0.148	0.028	0.020	0.010	1.183
Nappe 1	0.514	0.237	0.008	0.015	0.000	0.774
Sorting 1	0.464	0.534	0.071	0.001	0.001	1.072
Nappe 2	0.505	0.148	0.002	0.009	0.008	0.672
Sorting 2	0.472	0.000	0.026	0.011	0.003	0.511

[a] By axis and by node.

to an opposition between classes of the sorting and thus to the two kinds of data). These results are entirely satisfactory for users.

10.7.3 Decompositions of Inertia

Table 10.5 brings together the decomposition of inertia by axis and by node in the HMFA. Figure 10.9 illustrates the decomposition of the first axis according to the nodes.

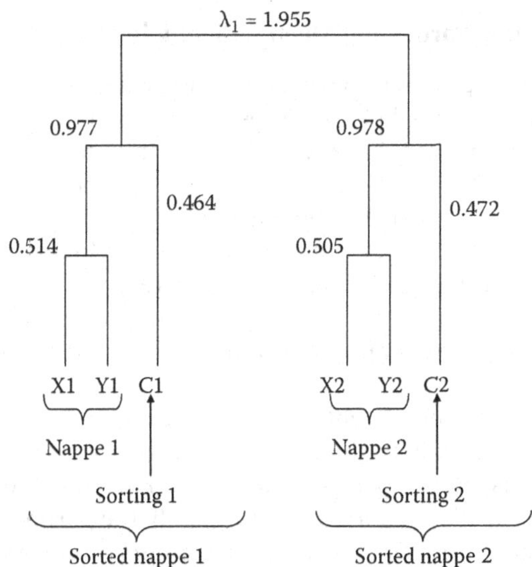

FIGURE 10.9
Two sorted nappes. HMFA. Decomposition of the inertia of the first axis according to the nodes of the hierarchy.

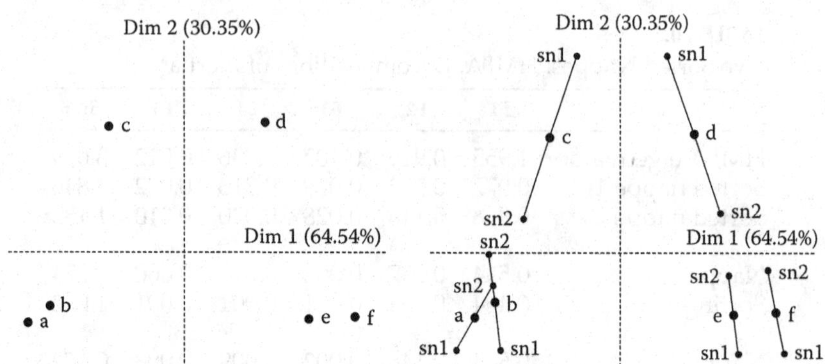

FIGURE 10.10
Two sorted nappes. HMFA. Representation of mean individuals (left) and partial individuals (right) associated with the root node (that is to say, associated with each sorted nappe, sn).

The first two axes make it possible to reconstitute the data almost perfectly ($(1.955 + .919)/3.029 = 94.89$ % of the inertia) and we can limit our comments to these two axes. The decomposition of the first eigenvalue according to the nodes of the hierarchy indicates balanced influence:

- Of the two sorted nappes one with the other (.977 *versus* .978)
- Of the nappe and the sorting within each sorted nappe (.514 *versus* .464 on the one hand and .505 *versus* .472)

10.7.4 Representing Partial and Mean Individuals

As the two sorted nappes have corresponding horizontal and vertical dimensions, it can easily be seen that the representation of individuals generated by the HMFA (see Figure 10.10) is much like a mean of these two nappes.

In particular, the deviation between the first two percentages of inertia is, in HMFA, intermediary to that which is in the two sorted nodes ($12.23\% < 34.199\% < 70.94\%$; the percentages of the sorted nappes can be seen in Figure 10.7; for example: $12.23\% = 54.17\% - 41.94\%$).

Another example: in the mean configuration, points *c* and *d* are closer to each other than on sorted nappe 2 and more distanced from each other than on sorted nappe 1.

In HMFA, for each individual, it is possible to represent as many partial points as there are nodes in the hierarchy. In practice, we begin by representing the partial points associated with the root node. If necessary, we go down in the hierarchy. In the practice of sorted napping, only the partial points associated with the root node are used: each one represents an individual in a sorted nappe.

The following remarks can be made about Figure 10.10. Along axis 1, the two partial points associated with one individual are very close; this is the case for all the individuals. This first dimension is a factor which is common to both groups of variables. A more detailed examination shows that individuals c and d are closer to each other on nappe 1 than on nappe 2, which can be seen both in the coordinates of the partial points along axis 1 and on the sorted nappes themselves. In even more detail, according to this dimension, which we outline in opposition with $\{a, b\}$ and $\{e, f\}$, a is more extreme than b only on nappe 1, which can be seen both on the plane generated by the HMFA and on the nappes themselves.

Along axis 2, the opposition between $\{c, d\}$ and the other individuals can be observed for the partial points of both nappes. This opposition is much more pronounced on the first nappe than on the second. This can be seen directly on the sorted nappes themselves.

Thus, the representation of partial points examined axis by axis does indeed lead to interpretations which can be clearly read in the data. However, it would seem that proportions are not respected between the axes:

- For nappe 1: The vertical dimension seems to be of greater importance (relative to the horizontal dimension) in the representation of partial points than in the sorted nappe.
- For nappe 2: The vertical dimension seems to be less important (relative to the horizontal dimension) in the partial representation than in the sorted nappe.

This visual impression is confirmed by confronting the variances of the two principal dimensions in the different representations (see Table 10.6): $.769 < 1.292$ and $19.383 > 6.229$.

We here observe a particularity of the superimposed representation in MFA described in Section 5.5 and seen in Section 9.3.2: when an MFA axis corresponds to different inertias depending on the group, these differences in inertia are amplified in the superimposed representation.

In practice, when analysing a set of sorted nappes, the superimposed representation is not described beyond the first partition. However, within the context of this methodological study, a superimposed representation made

TABLE 10.6
Two Sorted Nappes[a]

	F1	*F2*	*F1/F1*
MFA nappe 1	1.866	1.444	1.292
MFA nappe 2	1.956	0.314	6.229
HMFA partial cloud	1.955	0.920	2.126
HMFA partial cloud nappe 1	1.998	2.598	0.769
HMFA partial cloud nappe 2	2.000	0.103	19.383

[a] Variance of the first two factors in the sorted nappes and in the representation generated by the MFA (partial and mean clouds).

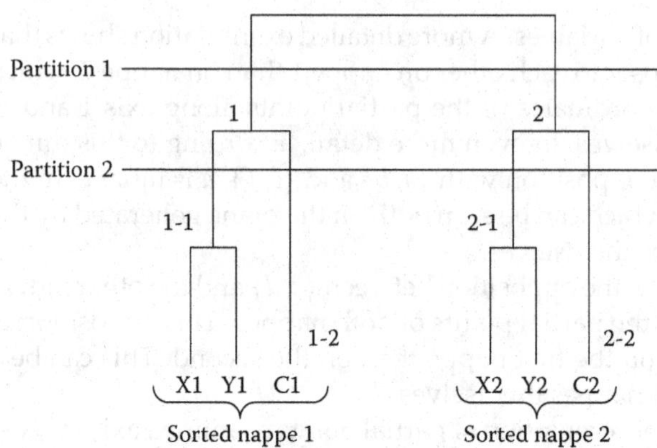

FIGURE 10.11
Two sorted nappes. HMFA. Numbering the nodes for the superimposed representation.

up of at least two levels of partition needs to be studied, particularly in the case of variables of different types (quantitative/qualitative).

In the following, each node is numbered indicating the path linking it to the root node of the tree (see Figure 10.11). The label of a partial point concatenates that of a given individual and node.

Figure 10.12 repeats Figure 10.10, adding the partial clouds associated with the partition 2.

Let us consider points *c* and *d* from the point of view of sorted nappe 1 (*c*-1 and *d*-1). When accounting for partition 2, each of these points now appears:

- From the point of view of the sorting alone (*c*-1-2 and *d*-1-2), in which case *c* and *d* lie in the same place
- From the point of view of the nappe (*c*-1-1 and *d*-1-1), in which case *c* and *d* are more distanced than on the sorted nappe (as the sorting brings together *c*-1 and *d*-1)

Let us consider points *c* and *d* from the point of view of sorted nappe 2 (*c*-2 and *d*-2). By differentiating between the nappe and the sorting, Figure 10.12 shows that points *c* and *d*:

- Are not characterised by axis 2 when seen from the perspective of sorting alone (*c*-2-2 and *d*-2-2), which clearly represents the noncharacterisation of these points by qualitative variable S2 (they do not belong to the same group unlike for S1).
- Are characterised by axis 2 when seen from the perspective of the nappe (*c*-2-1 and *d*-2-1); their coordinates are slightly lower than those of their counterparts on nappe 1 (*c*-1-1 and *d*-1-1), which can be seen on the nappes (see Figure 10.7).

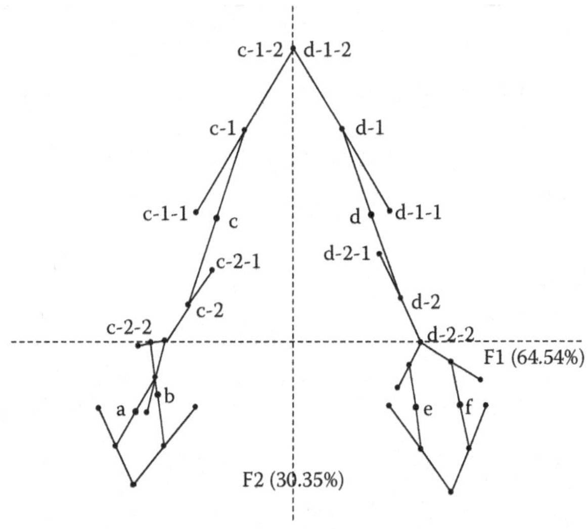

FIGURE 10.12
Two sorted nappes. HMFA. Superimposed representation of the partial clouds associated with the first two partitions (see Figure 10.11). Only the points related to individuals c and d are labelled.

Representation of Groups of Variables (see Figure 10.13)

In HMFA, a group of variables corresponds to a node of the tree structuring the variables. The Ng indicator (squared norm in $\mathbb{R}^{I^2}$), which specifies the group's dimensionality, appears in the first column of Table 10.7. This indicator shows:

- The dimensions (equal to 2 or 1) of the sortings (qualitative variables with three or two categories)
- The strongest dimensionality of nappe 1 (1.198 > 1.099) due to the greater relative importance of its second (vertical) dimension
- The dimensionalities of these sorted nappes, intermediate between that of the nappes and that of the sortings

For the two groups attached to the root node (sorted nappes 1 and 2), the representation of nodes in HMFA is exactly the same as that for the groups in MFA: the coordinates of the groups are interpreted both as a contribution and a relationship measurement. In this example, these coordinates are the inertias of rows 2 and 3 of Table 10.5. Figure 10.13 thus illustrates:

- The great importance of the first axis on both sorted nappes
- The importance of axis 2 greater in sorted nappe 1 (in which this axis corresponds to both kinds of data) than in sorted nappe 2 (in which this axis is not at all linked to the sorting)

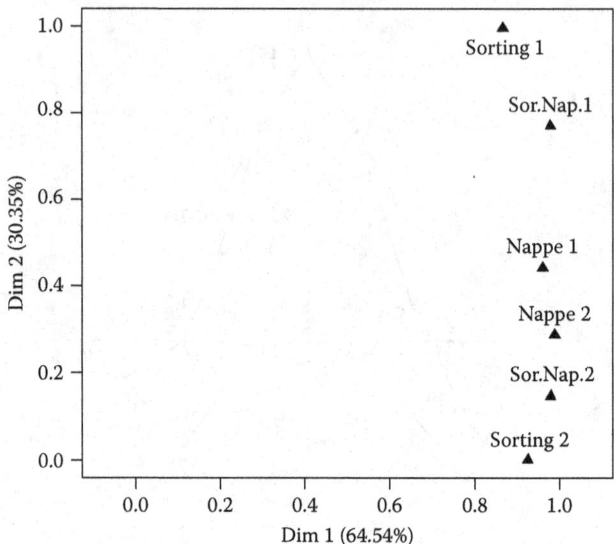

FIGURE 10.13
Two sorted nappes. HMFA. Representing the nodes (= groups of variables) in the relationship square.

For the other groups, the coordinates are interpreted only as the relationship measurement *Lg*. In particular, the *sorting* nodes, corresponding to a single qualitative variable, have a coordinate equal to ϕ^2 (between the variable and the factor). Thus, in the example:

TABLE 10.7
Two Sorted Nappes. HMFA[a]

Nodes	*Ng*	*F1*	*F2*	Plane(1,2)
Sorted Nappe 1	1.604	0.595	0.371	0.966
Sorted Nappe 2	1.026	0.931	0.021	0.953
Nappe 1	1.198	0.767	0.164	0.931
Sorting 1	2.000	0.374	0.497	0.871
Nappe 2	1.099	0.889	0.076	0.965
Sorting 2	1.000	0.853	0.000	0.853

[a] Squared norms (*Ng*) and qualities of representations of the nodes in $\mathbb{R}^{I^2}$.

- The first axis is an important dimension from both perspectives (nappe and sorting) for each judge.
- The second axis is a dimension of almost equal importance in both nappes; it is closely related to the sorting of judge 1 and not at all to that of judge 2.

This representation can be completed by a quality of representation indicator (see Table 10.7). In this simple example, all of the groups are well represented ($\cos^2 \geq .853$) by the first plane.

From this representation, the user will see a strong similarity between the nappes and a contrasting situation for the sortings (strong similarity according to one axis, strong dissimilarity according to the other). This situation is extreme in the example, but aside from these particular data, it is related to the general nature of the data. Indeed this is the advantage of this methodology which incorporates both quantitative aspects (with the possibility of nuances) and qualitative aspects (focussing on the essential).

10.8 HMFA in FactoMineR

Here we use the *Two sorted nappes* data (see Table 10.4). This presentation focusses on the specific aspects of the HMFA (compared with the MFA).

```
# Importation
> SorNap=read.table("SorNap.csv",header=T,sep=";",row.names=1)

# Checking the first two rows
> SorNap[1:2,]
  X1 Y1   S1 X2 Y2   S2
a 10 10 S1_1 10 10 S2_1
b 20 10 S1_1 10 20 S2_1
```

To define the hierarchy on the variables, we describe a sequence of nested partitions. The first partition described is the finest (= most detailed). Then, for each partition level, the groups defined by the previous partition are aggregated.

In the example of the two sorted nappes, the four groups of variables (nappe and sorting for each taster) are defined first. Following a procedure similar to that of MFA, we obtain four groups (see Figure 10.14, left), two of which are quantitative (each containing the two coordinates of a nappe, called `Nappe_1` and `Nappe_2`) and two of which are qualitative (each including a qualitative variable, called `Sorting_1` and `Sorting_2`). Once this has been done, the *Next level of the hierarchy* button opens a window (see Figure 10.14, right) to bring together the groups which have already been constructed. On the right

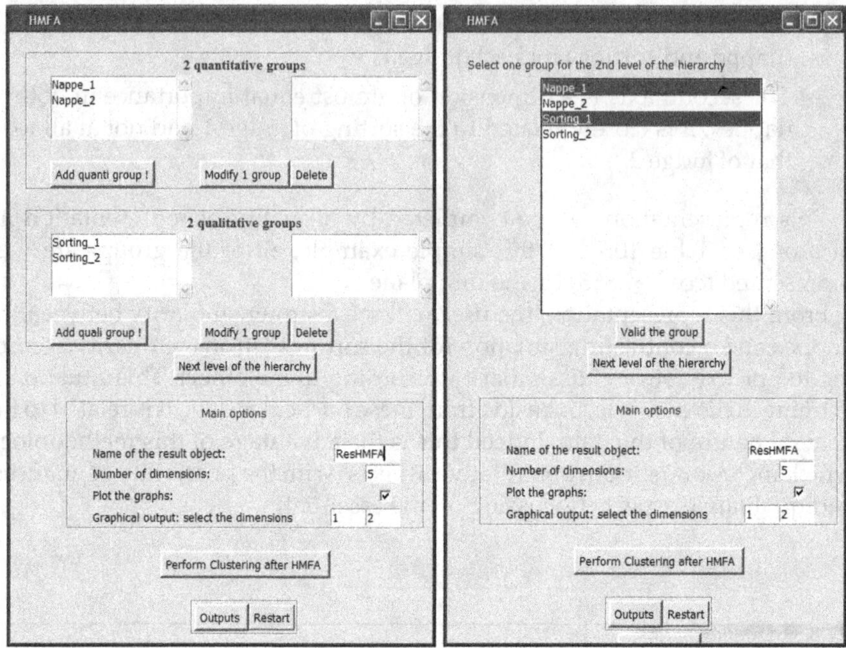

FIGURE 10.14
HMFA in **FactoMineR**. Windows for the finest partition (left) and the next partition (right).

of Figure 10.14, the two groups of sorted nappe 1 (`Nappe_1` and `Sorting_1`) are selected.

Remark
As in MFA where it is possible to include a single variable in several groups, in HMFA it is possible to include a group from a given hierarchical level in several groups of the next level.

Again here, it is possible to launch a clustering directly from the factors resulting from the MFA.

The command line for this HMFA is written:

```
> ResHMFA=HMFA(SorNap,type=c("c","n","c","n"),H=list
+ (c(2,1,2,1),c(2,2)),name.group=list(c("Nappe 1","Sorting 1",
+ "Nappe 2","Sorting 2"),c("Sor.Nap.1","Sor.Nap.2")))
```

File `SorNap` can be directly subjected to HMFA inasmuch as:

– The variables belonging to one group of the finest partition are adjacent.
– The groups of this partition brought together in the superior partition are adjacent.

The hierarchy is entered by the argument `H`, which defines a list containing as many terms as partitions. The first of these terms corresponds to the finest partition which, like in MFA, is described by the number of variables of each group (in the example, the first group contains the first two variables, the second only the third and so on). The next term describes the next partition, using the same principle applied to the groups of the previous partition. In the example, this second partition includes a first group which brings together the first two groups of the previous hierarchical level and a second group which brings together the last two.

The type of group is specified in the argument `type` which is endowed with a sequence of characters (as in MFA: `"c"` = unreduced quantitative group; `"s"` = standardised; `"n"` = qualitative group), each corresponding to a group of the finest partition.

It is possible to name each group, wherever it is in the hierarchical tree. These labels are brought together in a list (attributed to `name.group`) with the same structure as that which defined the hierarchy (attributed to `H`). In the example, the two groups of the second partition are called `Sor.Nap.1` and `Sor.Nap.2`.

Many graphs can be generated using the `plot.HMFA` function, for example, Figure 10.10:

```
> plot.HMFA(ResAFMH,choix="ind",invisible="quali",new.plot=
  TRUE,
+ cex=1.4)
```

The argument `choix` selects which elements to represent; `ind` represents the individuals and categories of the qualitative variables (as the centre of gravity of the individuals). The argument `invisible` is used to refine the selection: here, only the individuals are retained. This command generates three graphs including those in Figure 10.10. They are then all closed, except the graph of level 1 partial points, which becomes active.

Short labels are then attributed to the partial points using the following lines of code.

```
> text(ResHMFA$partial[[2]][,1:2,1],labels=rep("sn1",6),pos=3,
+ offset=.5,cex=1)
> text(ResHMFA$partial[[2]][,1:2,2],labels=rep("sn2",6),pos=3,
+ offset=.5,cex=1)
```

Users often want to present the results in a specific way. This is the case for Table 10.5 obtained using the following R code.

```
# Initialisation
> Tab10_5=matrix(nrow=7,ncol=6)
# Names of rows and columns
> row.names(Tab10_5)=c("Eigenvalues HMFA","Sor.Nap.1",
+ "Sor.Nap.2",
```

```
+ "Nappe 1","Sorting 1","Nappe 2","Sorting 2")
> colnames(Tab10_5)=c(paste("F",1:5,sep=""),"Sum")

# Eigenvalues of HMFA
> Tab10_5[1,1:5]=t(ResHMFA$eig[,1])
# Inertia of sorted nappes
> Tab10_5[2:3,1:5]=ResHMFA$group$coord[[2]][,]

# Contributions (%) of nappes
# (The ctr of the 2 coordinates are added together)
> quan_ctr=ResHMFA$quanti.var$contrib
> Tab10_5[4,1:5]=apply(quan_ctr[1:2,],MARGIN=2,FUN=sum)
> Tab10_5[6,1:5]=apply(quan_ctr[3:4,],MARGIN=2,FUN=sum)

# Contribution (%) of sortings
# (the categories' ctr are added together)
> qual_ctr=ResHMFA$quali.var$contrib
> Tab10_5[5,1:5]=apply(qual_ctr[1:3,],MARGIN=2,FUN=sum)
> Tab10_5[7,1:5]=apply(qual_ctr[4:5,],MARGIN=2,FUN=sum)

# Inertias are obtained by multiplying the contributions (%)
# by eigenvalues
> Tab10_5[4:7,1:5]=Tab10_5[4:7,1:5]%*%diag(ResHMFA$eig[,1])/100

# Rowise sum (last column)
Tab10_5[,6]=apply(Tab10_5[,1:5],MARGIN=1,FUN=sum)

# Edition with 3 places
> round(Tab10_5,3)
```

Table 10.6 brings together the results from the HMFA and the separate MFA of the nodes linked to the root node. Table 10.7 requires calulations in $\mathbb{R}^{I^2}$. The code for these two tables can be found below. As an exercise, readers might like to try to find it themselves.

```
# Table 10.6
# Initialisation
> Tab10_6=matrix(nrow=5,ncol=3)
# Names of rows and columns
> row.names(Tab10_6)=c("MFA nappe 1","MFA nappe 2","HMFA mean
+ cloud",
+ "HMFA partial cloud nappe 1","HMFA partial cloud nappe 2")
> colnames(Tab10_6)=c("F1","F2","F1/F2")

# Eigenvalues of separate MFA of sorted nappes

> Tab10_6[1,1:2]=resMFAnappe1$eig[1:2,1]
> Tab10_6[2,1:2]=resMFAnappe2$eig[1:2,1]
```

```
# HMFA. Eigevalues and then variance of partial clouds
# by HMFA dimension
> Tab10_6[3,1:2]=ResHMFA$eig[1:2,1]
> Tab10_6[4,1:2]=apply(ResHMFA$partial[[2]][,1:2,1],MARGIN=2,
+ FUN=var)*5/6
> Tab10_6[5,1:2]=apply(ResHMFA$partial[[2]][,1:2,2],MARGIN=2,
+ FUN=var)*5/6

> for(i in 1:5){Tab10_6[i,3]=Tab10_6[i,1]/Tab10_6[i,2]}
> round(Tab10_6,3)

# Table 10.7
# Initialisation
> Tab10_7=matrix(nrow=6,ncol=4)
# Names of rows and columns
> row.names(Tab10_7)=c("Sorted Nappe 1","Sorted Nappe 2",
+ "Nappe 1","Sorting 1","Nappe 2","Sorting 2")
> colnames(Tab10_7)=c("Ng","F1","F2","Plane(1,2)")

# Groups' norms before the last weighting of HMFA (Ng)
# are in the separate MFA of the sorted nappes
> Tab10_7[1,1]=sum(resMFAnappe1$eig[,1]^2)/resMFAnappe1$eig
+ [1,1]^2
> Tab10_7[2,1]=sum(resMFAnappe2$eig[,1]^2)/resMFAnappe2$eig
+ [1,1]^2
> Tab10_7[3:4,1]=diag(resMFAnappe1$group$Lg)[1:2]
> Tab10_7[5:6,1]=diag(resMFAnappe2$group$Lg)[1:2]

# Squared cosine of groups: squared projected length (in HMFA)
# divided by squared total length (Ng)
> for(i in 1:2){Tab10_7[1:2,i+1]=ResHMFA$group$coord[[2]]
+ [,i]^2/Tab10_7[1:2,1]}
> for(i in 1:2){Tab10_7[3:6,i+1]=ResHMFA$group$coord[[1]]
+ [,i]^2/Tab10_7[3:6,1]}
> Tab10_7[,4]=apply(Tab10_7[,2:3],MARGIN=1,FUN=sum)
> round(Tab10_7,3)
```

Below are the lines of code corresponding to the the HMFA as applied to the Orange Juice data.

```
# Importation and selection of required columns
# in orange data frame
> orange5=read.csv2("orange5.csv",header=T,row.names=1)
> orange=orange5[,c(3:17,19:114)]

# HMFA
> resHMFA=HMFA(orange,type=c("s","s","s"),H=list(c(8,7,96),
+ c(2,1)),name.group=list(c("Chemical","Sensory","Hedonic"),
+ c("Characterisation","Hedonic")))
```

```
# Figure 10.3
> plot.HMFA(resHMFA,choix="ind",invisible="quali",
+ new.plot=TRUE,cex=1.4)
# This command produces 3 graphs including those of Figure 10.3.
# They must be closed, except the one with level 1 partial points.
# The latter thus becomes active.
> text(resHMFA$partial[[2]][,1:2,1],labels=rep("c",6),pos=3,
+ offset=.5,cex=1)
> text(resHMFA$partial[[2]][,1:2,2],labels=rep("h",6),pos=3,
+ offset=.5,cex=1)

# Figure 10.4
> plot.HMFA(resHMFA,choix="ind",invisible="quali",
+ new.plot=TRUE,cex=1.4)
# This command produces 3 graphs including those of Figure 10.4.
# They must be closed, except the one with level 1 partial points.
# The latter thus becomes active.
> text(resHMFA$partial[[2]][,1:2,1],labels=rep("c",6),pos=3,
+ offset=.5,cex=1)
> text(resHMFA$partial[[2]][,1:2,2],labels=rep("h",6),pos=3,
+ offset=.5,cex=1)
> text(resHMFA$partial[[1]][,1:2,1],labels=rep("ch",6),pos=3,
+ offset=.5,cex=1)
> text(resHMFA$partial[[1]][,1:2,2],labels=rep("s",6),pos=3,
+ offset=.5,cex=1)
```

To obtain Figure 10.5, the analyses are conducted node by node and the factors stored. A PCA is conducted on the first two factors of the HMFA (which reconstructs the HMFA) by introducing the factors of by node analyses as supplementary.

```
# Analyses by node
> resacpchim=PCA(orange[,c(1:8)],graph=FALSE)
> resacpsenso=PCA(orange[,9:15],graph=FALSE)
> resacphedo=PCA(orange[,16:111],graph=FALSE)
> resMFAcaract=MFA(orange[,1:15],group=c(8,7),type=c("s","s"),
+ name.group=c("Chemical","Sensory"),graph=FALSE)

# Concatenation of the first two factors
# by node and from HMFA
> axpartHMFA=cbind(resHMFA$ind$coord[,1:2],
+ resacpchim$ind$coord[,1:2],resacpsenso$ind$coord[,1:2],
+ resMFAcaract$ind$coord[,1:2],resacphedo$ind$coord[,1:2])
# Labels of partial axes
> colnames(axpartHMFA)=c("F1_HMFA","F2_HMFA","F1_Ch","F2_Ch",
+ "F1_S","F2_S","F1_C","F2_C","F1_H","F2_H")
```

```
# PCA on HMFA factors as active, the factors by node
# being supplementary; the first HMFA factor is used twice in
# order to preserve the rank order of factors even if this
# PCA is standardised. Figure 10.5 is the variables graph.
> resPartHMFA=PCA(axpartHMFA[,c(1,1,2:10)],quanti.sup=c(4:11))
> plot(resPartHMFA,choix="var")
```

$$\mathcal{X}'\mathcal{X}u=\lambda u$$

11

Matrix Calculus and Euclidean Vector Space

This book uses several elements of algebra, specifically matrix calculus and the notion of spaces endowed with a distance (or a metric). To help users avoid having to read a general algebra book, we have grouped these elements together. These examples do not need to be read in any particular order, given that each one uses elements of the other.

11.1 Matrix Calculus

Definitions

A matrix is a set of numbers organised in a rectangular table. It is generally denoted by a capital letter (for example, X). Its terms (= elements = entries) are designated using indices, the first being that of the rows (for example, A_{12}, at the crossing of the first row and the second column).

The dimensions of a matrix are given in brackets (for example, matrix $A(n, p)$ has n rows and p columns). If $n = p$, the matrix is square.

In a symmetric square matrix $A_{ij} = A_{ji}$. In a diagonal matrix $A_{ij} = 0$ if $i \neq j$. A diagonal matrix with nonzero terms equal to 1 is called an *identity matrix*. A vector corresponds to a matrix with only one column (a column matrix).

Transposition

Let us consider a matrix denoted X with n rows and p columns. Its transpose (matrix), denoted X', is obtained by writing the values of the rows of X in the columns of X'. The matrix X' thus obtained possesses p rows and n columns.

Matrix Multiplication

Let us consider two matrices $X(n, p)$ and $Y(p, q)$ in which the number of columns of X is equal to the number of rows of Y. The multiplication of these two matrices, denoted XY, is a matrix, denoted A, with n rows and q columns, and for which the general term A_{ij}, at the crossing of row i and column j, is the scalar product between the vector containing the terms of the ith row of

X and the vector containing the terms of the jth column of Y. Thus

$$A_{ij} = \sum_k X_{ik} Y_{kj}.$$

The (usual) scalar product between two vectors u and v (denoted $\langle u, v \rangle$) is obtained by multiplying the corresponding matrices (also denoted u and v), the first having been transposed.

$$\langle u, v \rangle = u'v = v'u.$$

Matrix multiplication can therefore be seen as a juxtaposition of scalar products. In statistics, we encounter this when the scalar product matrix is calculated between individuals. Let $X(n, p)$ be the data table in which n individuals are described by p variables. The rows of X correspond to the transposition of the column vectors, each containing the data of a given individual. The matrix multiplication XX' is therefore the same as calculating the scalar products of each individual with each of the others. Indeed, the general term XX', at the crossing of row i and column l, is the scalar product between individuals i and l. On the diagonal, we therefore obtain the squared norms of the individuals.

The other matrix multiplication which is commonly used in statistics is $X'X$. It juxtaposes the scalar products between the vectors representing the variables. When the variables are centred (and standardised, respectively), the $X'X$ matrix contains the covariances (and correlation coefficients, respectively) between the variables (up to a coefficient n and only when the individuals have the same weights). When the individuals are not of the same weight, with these weights being positioned on the diagonal of diagonal matrix D, the covariance matrix is written $X'DX$ (the variables are centred).

In factorial analysis, clouds of points are often projected onto axes. When the coordinates of point i are in the ith row of matrix X, we calculate all of the coordinates of the projections of points i on unit vector u by conducting matrix multiplication Xu (which here is clearly interpreted as a juxtaposition of scalar products).

The matrix multiplication is associative:

$$ABC = (AB)C = A(BC).$$

To conduct matrix multiplication among three matrices, two of them (adjacent) are first multiplied together and then the result of this multiplication is multiplied by the third.

The transpose of a multiplication of two matrices is equal to the multiplication of the two transpose matrices, with their order in the multiplication being inverted: $(AB)' = B'A'$.

Trace of a Square Matrix

The trace of a square matrix A is the sum of its diagonal terms. It is denoted trace (A).

When applied to the matrix XX' of the scalar products between individuals, the trace is the sum of the squared norms of the individuals, that is to say, if all the individuals have a weight of 1, the total inertia of the cloud of individuals.

The *trace* operator has a remarkable property:

$$\text{trace}(AB) = \text{trace}(BA).$$

When applied to the covariance matrix between individuals, this property, expressed trace $(X'DX) = \text{trace}(XX'D)$, links the total inertia of the cloud of variables with that of the cloud of individuals.

Property: the trace of a square matrix is equal to the sum of its eigenvalues (defined below). This property is used in principal component analysis (PCA): the eigenvalues of $X'DX$ are the inertias projected on the axes. The trace of $X'DX$ is the total inertia which is thus decomposed on the factorial axes.

Matrix and Function, Orthogonal Matrix, Diagonalisation

Let us consider matrix $A(n, p)$ and vector u of $\mathbb{R}^p$ associated with a matrix, also denoted u, of dimensions (p,1). By multiplying A and u, we obtain vector $v = Au$ with n coordinates therefore belonging to $\mathbb{R}^n$. Matrix A thus corresponds to a function, which for all elements u of $\mathbb{R}^p$ attributes an element v of $\mathbb{R}^n$. Generally, we consider the function within a space (for example, from $\mathbb{R}^n$ to $\mathbb{R}^n$), therefore associated with a square matrix of dimension (n,n).

A square matrix $A(n, n)$ is said to be orthogonal if all its column vectors are orthogonal and of norm 1 (thus constituting an orthonormal base). A matrix such as this thus verifies $A'A = I_n$, where I_n is the identity matrix of size n. It can also be shown that this matrix A verifies $A'A = AA' = I_n$.

An orthogonal matrix corresponds to a function with a remarkable property: the norm of a vector remains unchanged by this function's transformation. Indeed,

$$\|Au\|^2 = u'A'Au = u'u = \|u\|^2.$$

A function such as this, which preserves the distance, is known as *isometry*. The most common example is rotation, which is used in Procrustes analysis.

Base Change. Let us consider an orthogonal matrix A and u_A a vector expressed in the base of the columns of A. To express u_A in the usual canonical base, expression denoted u, we write: $u = Au_A$. From here we deduce the equation which makes it possible to write u in the base (of the columns of) A: $u_A = A'u$.

Vector u is said to be the eigenvector of the square matrix A associated with the eigenvalue λ if it verifies $Au = \lambda u$.

Straight away, we can see that if u is an eigenvector, ku (k being a constant) is also an eigenvector. An eigenvector thus generates an eigendirection (or eigendimension) for which all of its vectors are eigenvectors (associated with the same eigenvalue). An eigendirection is represented by a unit vector (which leaves two possibilities between which the software chooses at random).

Geometric Interpretation. Let u be an eigenvector of matrix A associated with the eigenvalue λ. If the function associated with matrix A is applied to u, we obtain a vector collinear to u (with the λ ratio). The eigendirections are constant for function A and therefore are very particular for this function (and consequently for matrix A).

The procedure which finds the eigendimensions of matrix A (each represented by a unit vector) is called *diagonalisation*. The terminology 'the eigenvectors of $A(n, n)'$' designates a set of n unit eigenvectors, each associated with a distinct eigenvalue.

A symmetric matrix has the two following properties:

1. Two eigenvectors associated with two distinct eigenvalues are orthogonal (for the identity metric).
2. The eigenvalues are real numbers.

In PCA, in space $\mathbb{R}^K$ (endowed with the identity metric), the factorial axes are obtained from the eigenvectors of $X'DX$ (see Section 1.5.3). As this matrix is symmetrical, its eigenvectors are orthogonal (with one another), as are the factorial axes they define.

In space $\mathbb{R}^I$ (endowed with metric D of the weights of individuals), the vectors we are interested in, denoted v_s, are eigenvectors of $XX'D$. Thus

$$XX'Dv_s = \lambda_s v_s.$$

Matrix D is diagonal and contains only positive terms. We denote $D^{1/2}$ the diagonal matrix so that $D^{1/2}D^{1/2} = D$ (its terms are thus the root of those of D). By left-multiplying the two terms of the previous equation by $D^{1/2}$ we obtain

$$D^{1/2}XX'D^{1/2}D^{1/2}v_s = \lambda_s D^{1/2}v_s.$$

This illustrates that $D^{1/2}v_s$ is the eigenvector of $D^{1/2}XX'D^{1/2}$ associated with eigenvalue λ_s. As this matrix is symmetric, two eigenvectors associated with distinct eigenvalues are orthogonal for the usual matrix, thus

$$\left(D^{1/2}v_s\right)'\left(D^{1/2}v_t\right) = v_s'Dv_t,$$

which expresses that the axes generated by vectors v_s and v_t are orthogonal for metric D.

In factorial analysis, an eigenvalue corresponds to a projected inertia. It is therefore positive or zero. The condition of symmetry of A is not sufficient to ensure non-negative eigenvalues. In fact, when the diagonalised matrix can be written in the form $X'X$ or XX' (that is to say the multiplication of a matrix by its transpose), then the eigenvalues are positive or zero. Indeed, from

$$X'Xu = \lambda u,$$

it is possible to deduce

$$u'X'Xu = \lambda u'u$$

and therefore

$$\|Xu\|^2 = \lambda \, \|u\|^2 ,$$

which indicates that λ is positive or zero. Therefore, if u is unitary, λ is equal to the sum of squared coordinates of the projections of the rows of X on u (in PCA, this is the total inertia of the cloud of points for which the coordinates are in the rows of X).

11.2 Euclidean Vector Space

In factorial analysis, we work in vector spaces. The notions of distance, norm, projection and angle (and therefore scalar product) are essential to this type of analysis. Here, we show how these notions are related to one another and how they can be calculated.

A vector space (finite dimensional) in which a distance (= metric) is defined from a scalar product is said to be *Euclidean*. In factorial analyses, we always work in Euclidean spaces.

11.2.1 Vector Space Endowed with the Usual Distance

Two-Dimensional Space

In usual three-dimensional space ($\mathbb{R}^3$, that which is all around us), we are familiar with the notions of distance, length and angle. We start from this basic knowledge, first reasoning on the plane ($\mathbb{R}^2$) to keep things simple, and we conduct simple calculations from two points A and B (see Figure 11.1). By considering $\mathbb{R}^2$ as a vector space, A and B are therefore vectors, connecting the origin with points A and B (here we recognise the point of view of the variables in PCA). A as a vector is sometimes denoted $\vec{A}$ or even $\overrightarrow{OA}$. To simplify notation, we use the same letter A to designate the point, the vector and the matrix (with one column) bringing together the coordinates of A (x_a and y_a).

Pythagoras' theorem is used to calculate the squared distance between A and B.

$$d^2(A, B) = (x_a - x_b)^2 + (y_a - y_b)^2 .$$

The length of the vector $\overrightarrow{OA}$, also known as its norm, denoted $\|\overrightarrow{OA}\|$ or simply $\|A\|$, is calculated as the distance between O and A thus

$$\|\overrightarrow{OA}\|^2 = \|A\|^2 = d^2(O, A) .$$

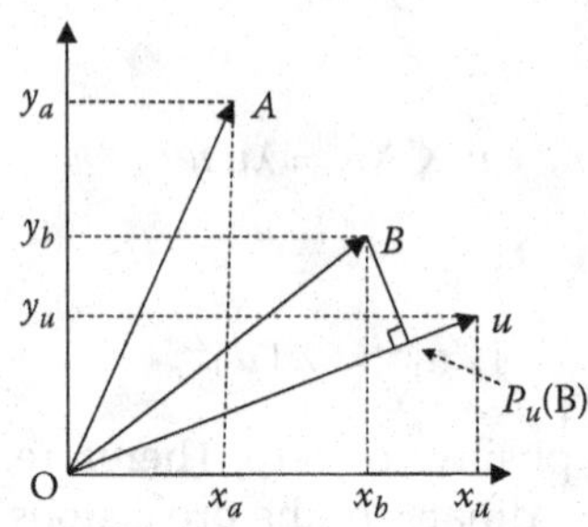

FIGURE 11.1
A few notations in $\mathbb{R}^2$.

At this level, we retain the fact that the notions of distance and norm are connected.

Distance is also related to the notion of projection. The projection of a point B on a line D is the point of the line D which is closest to B. In $\mathbb{R}^2$, to obtain the coordinate of this projection, the scalar product is calculated between B and a unit vector of D (denoted u, of coordinates x_u and y_u). $P_u(B)$ is the projection of B on u. The length of this projection is worth (where $\langle B, u\rangle$ is the scalar product between B and u):

$$\| P_u(B)\| = \langle B, u\rangle = |x_b x_u + y_b y_u|.$$

The notions of scalar product and norm are connected; by projecting a vector onto itself, it remains unchanged. Thus, for a vector v (of coordinates x_v and y_v):

$$\|v\|^2 = \langle v, v\rangle = x_v^2 + y_v^2.$$

Finally, the notion of scalar product is connected to the notion of angle as follows. Let u and v be two vectors forming an angle θ. The cosine of angle θ is obtained by projecting one of these two vectors on the other after having standardised them. Thus

$$\cos\theta = \langle \frac{u}{\|u\|}, \frac{v}{\|v\|}\rangle.$$

Note: Two vectors are said to be orthogonal if their scalar product is zero.

Summary. From the definition of scalar product, we define the norm: $\|u\|^2 = \langle u, u\rangle$. From the norm, we define the distance: $d^2(u, v) = \|u - v\|^2$.

Vector Space with *n* Dimensions

The scalar product previously defined in $\mathbb{R}^2$ can easily be generalised in $\mathbb{R}^n$. Therefore, denoting $\{u_i; i = 1, n\}$ the coordinates of u and $\{v_i; i = 1, n\}$ the coordinates of v:

$$\langle u, v\rangle = u_1 v_1 + u_2 v_2 + \cdots = \sum_{i=1}^{i=n} u_i v_i.$$

In matrix notations, retaining the notation u (and v, respectively) to designate the column matrix (with n rows and one column) bringing together the coordinates of u (and v, respectively), the scalar product between u and v is written (denoting u' the transpose matrix of u):

$$\langle u, v\rangle = u'v = v'u.$$

The orthogonality between two vectors is thus expressed: $u'v = 0$, and in addition: $\|u\|^2 = u'u$. A unit vector therefore satisfies $u'u = 1$.

11.2.2 Euclidean Space Endowed with a Diagonal Metric

Up until now, in the usual distance, the same weight 1 is attributed to each dimension within the space. In data analysis, we might want to attribute a weight other than 1 but identical for all of the dimensions, or a different weight for each dimension. The most common case is the cloud of variables in PCA.

This cloud evolves within a space with I dimensions, denoted $\mathbb{R}^I$, with each dimension associated with one individual. If individual i is endowed with weight p_i, then this same weight p_i must be attributed to dimension i when calculating distance in $\mathbb{R}^I$. Because, for simplicity's sake, we impose a total weight of 1 to p_i, these weights p_i are never equal to 1 and therefore, even in cases where the individuals have the same weights, the usual distance needs to be altered (slightly).

Let us look back at the general case of the space with n dimensions ($\mathbb{R}^n$) retaining index i for the dimensions. Let p_i be the weight attributed to dimension i. These weights are organised on the diagonal of matrix M, a matrix containing zeros elsewhere (matrix M is therefore diagonal).

To show that we use the weights organised within matrix M, this letter is mentioned in the notations of scalar product, norm and distance. Thus

$$\langle u, v\rangle_M = \sum_i p_i u_i v_i = u'Mv.$$

Hence the norm:

$$\|u\|_M^2 = \langle u, u\rangle_M = \sum_i p_i u_i^2 = u'Mu.$$

Hence the distance between (the extremities of the vectors) u and v:

$$d_M^2(u, v) = \|u - v\|_M^2 = \sum_i p_i (u_i - v_i)^2 .$$

Specific Cases. When all of the weights p_i are equal to 1, we obtain the usual (Euclidean) distance. Matrix M is then the identity matrix, hence the name *identity metric*. When all of the weights are equal to a constant c (and not 1; the usual case in space $\mathbb{R}^I$ of variables in PCA with $c = 1/I$), this is also referred to as the *identity metric*.

TABLE 11.1
Visualisation. Raw Data

	u_1	u_2		Dim1	Dim2		a	b	c	d
a	1	1	Dim1	4	0	a				
b	2	1	Dim2	0	1/4	b	2			
c	2	2				c	$\sqrt{17/2}$	1/2		
d	1	2				d	1/2	$\sqrt{17/2}$	1/2	
(a) Data			(b) Metric			(c) Distances				

In data analysis, the diagonal metrics have an essential role as they are easy to interpret; this is the same as attributing a weight to each dimension of the space. However, it is also possible to define a scalar product, and thus a metric, from a nondiagonal matrix M. This option is mentioned merely as an aside as it is not used in this work.

11.2.3 Visualising a Cloud in a Space Endowed with a Metric Different from the Identity

Visually, we can only read with the identity metric. Let us consider the data (named *Visualisation*) from Table 11.1 represented in Figure 11.2 (left). The four points $\{a, b, c, d\}$ form a square. However, this is a false impression if, for example, we use the metric attributing weight 4 to dimension 1 and 1–4 to dimension 2 (see distances in Table 11.1).

With this metric, vectors u_1 and u_2 which were used to construct the graph, are not unitary ($\|u_1\| = 2\|u_2\| = 1/2$). On the same axes, let us consider unit vectors e_1 and e_2. In Table 11.2, the data are expressed in base $\{e_1, e_2\}$ and the distances between the four points are calculated from these new data and with the identity metric.

The same distances are obtained as for Table 11.1. This time, the graph (see Figure 11.2 right) illustrates the distances correctly.

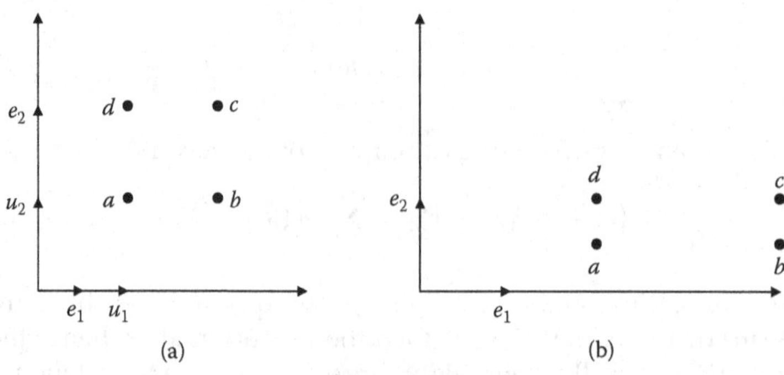

FIGURE 11.2
Visualisation. Representation of the data in Table 11.1 (left) and Table 11.2 (right).

TABLE 11.2
Visualisation. Data Expressed in an Orthonormal Base

	u_1	u_2		Dim1	Dim2		a	b	c	d
a	2	1/2	Dim1	1	0	a				
b	4	1/2	Dim2	0	1	b	2			
c	4	1				c	$\sqrt{17/2}$	1/2		
d	2	1				d	1/2	$\sqrt{17/2}$	1/2	
(a) Data			(b) Metric			(c) Distances				

This property is constantly used in factorial analysis: when using an Euclidean metric which is different from the identity, the data must be expressed in an orthonormal base; the representation generated is read correctly with the usual distance.

$$\langle u, v \rangle = u'v = v'u$$

Bibliography

Bécue M. & Pagès J. (2003). A principal axes method for comparing contingency tables: MFACT. *Computational Statistics and Data Analysis*, **45**(3), 481–503.

Bécue M. & Pagès J. (2008). Analysis of a mixture of quantitative, categorical and frequency data through an extension of multiple factor analysis: Application to survey data. *Computational Statistics and Data Analysis*, **53**(6), 3255–3268.

Bry X. (1996). *Analyses factorielles multiples*. Economica, Paris.

Cadoret M., Lê S. & Pagès J. (2011). Statistical analysis of hierarchical sorting data. *Journal of Sensory Studies*, **26**(2), 86–105.

Escofier B. & Pagès J. (1982a). Comparaison de groupes de variables: 2e partie: un exemple d'application. *Rapport de recherche INRIA*, (165).

Escofier B. & Pagès J. (1982b). Comparaison de groupes de variables définies sur le même ensemble d'individus. *Rapport de recherche INRIA*, **149**.

Escofier B. & Pagès J. (1983). Méthode pour l'analyse de plusieurs groupes de variables. *Revue de Statistique Appliquée*, **XXXI**(2), 43–59.

Escofier B. & Pagès J. (1984). L'analyse factorielle multiple. *Cahiers du BURO (Bureau Universitaire de Recherche Opérationnelle)*, **42**, 3–68.

Escofier B. & Pagès J. (2008). *Analyses factorielles simples et multiples; objectifs, méthodes et interprétation*. Dunod, Paris, 4e ed.

Fox J., with contributions from Liviu Andronic, Ash M., Boye T., Calza S., Chang A., Grosjean P., Heiberger R., Kerns G.J., Lancelot R., Lesnoff M., Ligges U., Messad S., Maechler M., Muenchen R., Murdoch D., Neuwirth E., Putler D., Ripley B., Ristic M., & Wolf. P. (2009). *Rcmdr: R Commander*. `http://CRAN.R-project.org/package=Rcmdr`. R package version 1.5-4.

Husson F. & Lê S. (2009). *SensoMineR: Sensory Data Analysis with R*. `http://CRAN.R-project.org/package=SensoMineR`. R package version 1.10.

Husson F. & Pagès J. (2006). Indscal model: Geometrical interpretation and methodology. *Computational Statistics and Data Analysis*, **50**(2), 358–378.

Husson F., Josse J., Lê S. & Mazet J. (2009). *FactoMineR: Factor Analysis and Data Mining with R.* `http://CRAN.R-project.org/package=FactoMineR`. R package version 1.12.

Lavit C. (1976). *Analyse conjointe de tableaux quantitatifs: Méthode et programmes.* Masson, Paris.

Lê S. & Husson F. (2008). SensoMineR: A package for sensory data analysis. *Journal of Sensory Studies*, **23**, 14–25.

Lê S., Josse J. & Husson F. (2008). FactoMineR: An R package for multivariate analysis. *Journal of Statistical Software*, **25**, 1–18.

Le Barzic J.F., Dazy F., Lavallard F. & Saporta G. (1996). *L'analyse des données évolutives. – Méthodes et Applications.* Technip, Paris.

Lê Dien S. & Pagès J. (2003a). Analyse factorielle multiple hiérarchique. *Revue de Statistique Appliquée*, **LI**(2), 47–73.

Lê Dien S. & Pagès J. (2003b). Hierarchical multiple factor analysis: Application to the comparison of sensory profiles. *Food Quality and Preference*, **14**, 397–403.

Lebart L., Piron M. & Morineau A. (2006). *Statistique exploratoire multidimensionnelle.* Dunod, Paris, 4e ed.

Morand E. & Pagès J. (2006). Procrustes multiple factor analysis to analyse the overall perception of food products. *Food Quality and Preference*, **17**, 36–42.

Morand E. & Pagès J. (2007). L'analyse factorielle multiple procustéenne. *Journal de la Société Française de Statistique*, **148**(2), 65–67.

Pagès J. (1996). Eléments de comparaison entre l'analyse factorielle multiple et la méthode STATIS. *Revue de Statistique Appliquée*, **XLIV**(4), 81–95.

Pagès J. (2002). Analyse factorielle multiple appliquée aux variables qualitatives et aux données mixtes. *Revue de Statistique Appliquée*, **L**(4), 5–37.

Pagès J. (2004). Analyse factorielle de données mixtes. *Revue de Statistique Appliquée*, **LII**(4), 93–111.

Pagès J. (2005a). Analyse factorielle multiple et analyse procustéenne. *Revue de Statistique Appliquée*, **LIII**(4), 61–68.

Pagès J. (2005b). Collection and analysis of perceived product inter-distances using multiple factor analysis; Application to the study of ten white wines from the Loire Valley. *Food Quality and Preference*, **16**, 642–649.

Pagès J. & Camiz S. (2008). Analyse factorielle multiple de données mixtes: application à la comparaison de deux codages. *Revue de Modulad*, **38**, 178–183.

Pagès J. & Husson F. (2005). Multiple factor analysis with confidence ellipses: A methodology to study the relationships between sensory and instrumental data. *J. Chemometrics.*, **19**, 1–7.

Pagès J. & Tenenhaus M. (2001). Multiple factor analysis combined with path modelling. Application to the analysis of relationships between physicochemical variables, sensory profiles and hedonic judgements. *Chemometrics and Intelligent Laboratory Systems*, **58**, 261–273.

Pagès J. & Tenenhaus M. (2002). Analyse factorielle multiple et approche PLS. *Revue de Statistique Appliquée*, **L**(1), 5–33.

Pagès J., Cadoret M. & Lê S. (2010). The sorted napping: A new holistic approach in sensory evaluation. *Journal of Sensory Studies*, **25**(5), 637–658.

R Development Core Team (2008). *R: A Language and Environment for Statistical Computing*. R Foundation for Statistical Computing, Vienna, Austria. `http://www.R-project.org`. ISBN 3-900051-07-0.

Saporta G. (2006). *Probabilités, analyse des données et statistique*. Technip, Paris, 2e ed.

Index

D

E

F

G

H

Printed in the United States
by Baker & Taylor Publisher Services